Springer Theses

Recognizing Outstanding Ph.D. Research

Aims and Scope

The series "Springer Theses" brings together a selection of the very best Ph.D. theses from around the world and across the physical sciences. Nominated and endorsed by two recognized specialists, each published volume has been selected for its scientific excellence and the high impact of its contents for the pertinent field of research. For greater accessibility to non-specialists, the published versions include an extended introduction, as well as a foreword by the student's supervisor explaining the special relevance of the work for the field. As a whole, the series will provide a valuable resource both for newcomers to the research fields described, and for other scientists seeking detailed background information on special questions. Finally, it provides an accredited documentation of the valuable contributions made by today's younger generation of scientists.

Theses are accepted into the series by invited nomination only and must fulfill all of the following criteria

- They must be written in good English.
- The topic of should fall within the confines of Chemistry, Physics and related interdisciplinary fields such as Materials, Nanoscience, Chemical Engineering, Complex Systems and Biophysics.
- The work reported in the thesis must represent a significant scientific advance.
- If the thesis includes previously published material, permission to reproduce this must be gained from the respective copyright holder.
- They must have been examined and passed during the 12 months prior to nomination.
- Each thesis should include a foreword by the supervisor outlining the significance of its content.
- The theses should have a clearly defined structure including an introduction accessible to scientists not expert in that particular field.

Massimo Baroncini

Design, Synthesis and Characterization of new Supramolecular Architectures

Doctoral Thesis accepted by
University of Bologna, Italy

 Springer

Author
Dr. Massimo Baroncini
Department of Chemistry
  "G. Ciamician"
University of Bologna
Via Selmi 2
40126 Bologna
Italy
e-mail: massimo.baroncini@fastwebnet.it

Supervisor
Prof. Margherita Venturi
Department of Chemistry
  "G. Ciamician"
University of Bologna
Via Selmi 2
40126 Bologna
Italy
e-mail: margherita.venturi@unibo.it

ISSN 2190-5053 e-ISSN 2190-5061

ISBN 978-3-642-26791-8 ISBN 978-3-642-19285-2 (eBook)

DOI 10.1007/978-3-642-19285-2

Springer Heidelberg Dordrecht London New York

Parts of this thesis have been published in the following journal articles

Silvi S, Arduini A, Pochini A, Secchi A, Tomasulo M, Raymo FM, Baroncini M, Credi A et al (2007) Reproduced with permission. J Am Chem Soc 129(44): 13378–13379

Ronconi MC, Stoddart JF, Balzani V, Baroncini M, Ceroni P, Giansante C, Venturi M et al (2009) Reproduced with permission. Chem Eur J 14(27):8365–8373

Amelia M, Baroncini M, Credi A et al (2009) Reproduced with permission. Angew Chem Int Ed 47(33):6240–6243

Giansante C, Mazzanti A, Baroncini M, Ceroni P, Venturi M, Klarner FG,Vogtle FJ et al (2009) Reproduced with permission. Org Chem 74(19):7335–7343

Semeraro M, Arduini A, Baroncini M, Battelli R, Credi A, Venturi M, Pochini A, Secchi A, Silvi S et al (2010) Reproduced with permission. Chem Eur J 16(11):3467–3475

Zappacosta R, Semeraro M, Baroncini M, Silvi S, Aschi M, Credi A, Fontana A et al (2010) Reproduced with permission. Small 6(8):952–959

Baroncini M, Silvi S, Venturi M, Credi A et al (2010) Reproduced with permission. Chem Eur J 16(38):11580–11587

Supervisor's Foreword

The design, synthesis and study of new supramolecular architectures are central topics for the development of new materials and technologies in the rapidly expanding field of the nanosciences. The number of publications dealing with the preparation and characterization of new supramolecular systems is continuously growing. Since the introduction of the concept of supramolecular chemistry by J. M. Lehn and the first pioneering papers there has been a huge expansion in the varieties of structures and properties developed by different research groups around the world. This thesis deals with some of the most advanced and studied supramolecular systems known today with the general aim of studying their properties to develop better ones. In the first chapters two studies are reported concerning the synthesis and characterization of dendrimers, large tree-like supramolecular structures, incorporating photo- and electro-active units. These systems are able to complex a large number of smaller molecules in the cavities present in their structure. These dendrimers behave, therefore, as molecular scaffolds and owing to the photo- and electro-active units incorporated in their structure the complexation process can be controlled by external stimuli. Furthermore, the presence of redox-active units enable the accumulation in their structure of a large number of electrons at a fixed potential value. Much is expected from these kind of dendritic structures in the most different fields of application like, just to mention the most investigated ones, controlled release of drugs, hybrid approaches to molecular electronics and light to energy conversion. The other part of the thesis deals with the study of supramolecular systems called molecular machines because of their capability to mimic some of the functions of the machines of the macroscopic world. Specifically, in three different works molecular machines able to change their behaviour as a result of different chemical or photochemical external stimuli are reported. These molecular machines are designed to be controlled by predefined external inputs adding a new level of control in their basic operations. The field of molecular machines is surely a growing one, although there are still few applications; it is, however, only a matter of time and resources before many practical results will be reached. In the last chapter of the thesis it is shown that simple everyday molecules could aptly work as logic units in an hypothetical chemical computer able to apply the simpler rules of Boolean algebra. To conclude it is apparent that this thesis spans a large number of open problems in the nanosciences, dealing with a wide range of the most advanced applications of supramolecular systems.

Bologna, March 2011   Prof. Margherita Venturi

Acknowledgments

The road that led to the completion of this book was difficult, however, I enjoyed the support of many wonderful people who guided and helped me along the way. The most influential person was Professor Margherita Venturi whose insight, honest criticism and invaluable suggestions helped to mold the manuscript into its current form. I am also especially indebted to Professor Alberto Credi and Professor Paola Ceroni for their vast experience and continual support. I also want to thank all my collegues and coworkers: Serena Silvi, Carlo Giansante, Giacomo Bergamini, Matteo Amelia, Monica Semeraro, Enrico Marchi and all the past and present members of the group of Photochemistry and Supramolecular Chemistry of the University of Bologna. Last but not least I want to express my gratitude to Professor Vincenzo Balzani for his continuous inspiration and support.

Bologna, March 2011   Massimo Baroncini

Contents

Chapter 1
Introduction

1.1 Supramolecular Chemistry

Supramolecular chemistry is a highly interdisciplinary field that has developed astonishingly rapidly in the last four decades [1–3]. In the late 1980s, following the award of the 1987 Nobel Prize to Pedersen, Cram, and Lehn, there was a sudden increase of interest in supramolecular chemistry, a highly interdisciplinary field based on concepts such as molecular recognition, preorganization, and self-assembling.

The classical definition of supramolecular chemistry is that given by J.-M. Lehn, namely "the chemistry beyond the molecule, bearing on organized entities of higher complexity that result from the association of two or more chemical species held together by intermolecular forces" [4].

There is, however, a problem with this definition. With supramolecular chemistry there has been a change in focus from molecules to molecular assemblies or multicomponent systems. According to the original definition, however, when the components of a chemical system are linked by covalent bonds, the system should not be considered a supramolecular species, but a molecule.

Although the classical definition of supramolecular chemistry as "the chemistry beyond the molecule" is quite useful in general, from a functional viewpoint the distinction between what is molecular and what is supramolecular can be better based on the degree of intercomponent electronic interactions [1, 5–8].

This concept is illustrated, for example, in Fig. 1.1 [9]. In the case of a photochemical stimulation, a system A $\sim$ B, consisting of two units ($\sim$ indicates any type of "bond" that keeps the units together), can be defined a supramolecular species if light absorption leads to excited states that are substantially localized on either A or B, or causes an electron transfer from A to B (or viceversa). By contrast, when the excited states are substantially delocalized on the entire system, the species can be better considered as a large molecule. Similarly (Fig. 1.1), oxidation and reduction of a supramolecular species can substantially be described

M. Baroncini, *Design, Synthesis and Characterization of new Supramolecular Architectures*, Springer Theses, DOI: 10.1007/978-3-642-19285-2_1,
© Springer-Verlag Berlin Heidelberg 2011

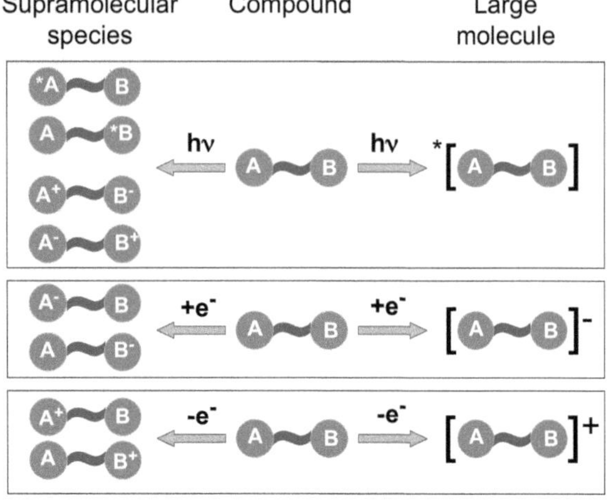

Fig. 1.1 Schematic representation of the difference between a supramolecular system and a large molecule based on the effects caused by a photon or an electron input (for more details, see text)

as oxidation and reduction of specific units, whereas oxidation and reduction of a large molecule leads to species where the hole or the electron are delocalized on the entire system. In more general terms, when the interaction energy between units is small compared to the other relevant energy parameters, a system can be considered a supramolecular species, regardless of the nature of the bonds that link the units. It should be noted that the properties of each component of a supramolecular species, i.e. of an assembly of weakly interacting molecular components, can be known from the study of the isolated components or of suitable model molecules.

1.2 Bottom-Up Approach to Nanotechnology

Nanotechnology is an experimental field of applied science and technology that has as its goal the realization of systems and devices for transforming matter, energy, and information, based on nanometer-scale components with precisely defined molecular features. The term nanotechnology has also been used more broadly to refer to techniques that produce or measure features less than 100 nm in size. Two main approaches are used in nanotechnology: one is a "bottom-up" approach where materials and devices are built from smaller (molecular) components which assemble themselves chemically using principles such as molecular recognition; the other being a "top-down" approach where they are synthesized or constructed from larger entities through an externally-controlled process.

The bottom-up approach to nanotechnology is relatively new. Until a few decades ago, in fact, nanotechnology was not considered an obtainable target by physicists [10]. The dominant idea, derived from quantum theory [11], was that atoms are fuzzy entities that "must no longer be regarded as identifiable individuals" [12], and "form a world of potentialities or possibilities rather than one of things or facts" [13]. From the point of view of quantum theory, molecular structure is not an intrinsic property, but a metaphor [14]. Such ideas, of course, were never shared by chemists who long before had established that atoms are material and reliable building blocks for constructing molecules and that molecules have well defined sizes and shapes [15].

The idea that atoms could be used to construct nanoscale machines was first raised by Feynman *"There is plenty of room at the bottom"* [16, 17]. The key sentence of Feynman's talk was: *"The principles of physics do not speak against the possibility of maneuvering things atom by atom"*. As we will see below, however, chemists do not believe in the possibility of realizing an atom-by-atom approach to nanostructures.

In the framework of research on supramolecular chemistry the idea began to arise in a few laboratories that molecules are much more convenient building blocks than atoms for construction of nanoscale devices and machines. The main foundations of this idea were: (a) molecules are stable species, whereas atoms are difficult to handle; (b) Nature starts from molecules, not from atoms, to construct the great number and variety of nanodevices and nanomachines that sustain life (vide infra); (c) most laboratory chemical processes deal with molecules, not with atoms; (d) molecules are objects that already have distinct shapes and carry device-related properties (e.g. properties that can be manipulated by photochemical and electrochemical inputs); and (e) molecules can self-assemble or can be connected to make larger structures.

In the following years supramolecular chemistry grew very rapidly [1–3] and it became clear that the supramolecular "bottom-up" approach opens virtually unlimited possibilities concerning design and construction of artificial molecular-level devices and machines. It also became increasingly evident that such an approach can make an invaluable contribution to a better understanding of molecular-level aspects of the extremely complicated devices and machines that are responsible for biological processes [18].

It should not be forgotten that the development of the supramolecular bottom-up approach towards the construction of nanodevices and nanomachines was made possible by the large amount of knowledge gained in other fields of chemistry. Particularly important in this regard have been the contributions made by organic synthesis, which supplied a variety of building blocks, and by photochemistry [1], which afforded a means of investigating the early examples of molecular-level devices and machines (e.g. light-controlled molecular-level tweezers [19], triads for vectorial charge separation [20], and light-harvesting antennae [21]).

It should also be recalled that in the last few years the concept of molecules as nanoscale objects with their own shape, size and properties has been confirmed by new, very powerful techniques, such as single-molecule fluorescence spectroscopy and the various types of probe microscopy, capable of "seeing" [22] or "manipulating" [23] single molecules. It has been possible, for example, to make ordered arrays of molecules (e.g. to write words [24] and numbers [25] by aligning single molecules in the desired pattern) and even to investigate bimolecular chemical reactions at the single molecule level [26].

References

1. Balzani V, Scandola F (1991) Supramolecular photochemistry. Horwood, Chichester
2. Vögtle F (1991) Supramolecular chemistry. An introduction. Wiley, Chichester
3. Whitesides GM, Mathias JP, Seto CT (1991) Science 254:1312
4. Lehn J-M (1995) Supramolecular chemistry: concepts and perspectives. VCH, Weinheim
5. Schneider HJ, Yatsimirsky A (2000) Principles and methods in supramolecular chemistry. Wiley, Chichester
6. Steed JW, Atwood JL (2000) Supramolecular chemistry. Wiley, Chichester
7. Lehn J-M (1993) In: Kisakürek MV (ed) Organic chemistry: its language and its state of the art. VCH, Weinheim, p 77
8. Kaifer AM, Gómez-Kaifer M (1999) Supramolecular electrochemistry. Wiley-VCH, Weinheim
9. Balzani V, Scandola F (1996) In: Atwood JL, Davies JED, Macnicol DD, Vögtle F (eds) Comprehensive supramolecular chemistry, vol 10. Pergamon Press, Oxford, p 687
10. Rouvrai D (2000) Chem Brit 36:46
11. Dirac PAM (1929) Proc R Soc Lond A 123:714
12. Schrödinger E (1951) Science and humanism: physics in our time. Cambridge University Press, Cambridge, p 27
13. Heisenberg W (1958) Physics and philosophy. Harper and Row, New York, p 186
14. Woolley RG (1985) J Chem Ed 62:1082
15. Atkins PW (1982) Physical chemistry, 2nd edn. Oxford University Press, Oxford Ch. 1
16. Feynman RP (1960) Eng Sci 23:22
17. Feynman RP (1960) Saturday Rev 43:45. http://www.its.caltech.edu/~feynman
18. Cramer F (1993) Chaos and order. The complex structure of living system. VCH, Weinheim
19. Shinkai S, Nakaji T, Ogawa T, Shigematsu K, Manabe O (1981) J Am Chem Soc 103:111
20. Seta P, Bienvenue E, Moore AL, Mathis P, Bensasson RV, Liddel P, Pessiky PJ, Joy A, Moore TA, Gust D (1985) Nature 316:653
21. Alpha B, Balzani V, Lehn J-M, Perathoner S, Sabbatici N (1987) Angew Chem Int Ed Engl 26:1266
22. Zander C, Enderlein J, Keller RA (eds) (2002) Single molecule detection in solution. Wiley-VCH, Weinheim
23. Crommie MF (2005) Science 309:1501
24. Schulz W (2000) Chem Eng News 78:41
25. Hla S-W, Meyer G, Rieder K-H (2001) ChemPhysChem 2:361
26. Christ T, Kulzer F, Bordat P, Basché T (2001) Angew Chem Int Ed 40:4192

Chapter 2
Supramolecular Architectures

2.1 Dendrimers

Dendrimers [1, 2] are globular size, monodisperse macromolecules in which all bonds emerge radially from a central focal point or core with a regular branching pattern and with repeating units constituting the branching points. The term *dendrimer* refers to its characteristic tree-like structure and it derives from the Greek word *dendron* (tree) and *meros* (part). From a topological viewpoint, dendrimers contain three different regions: core, branches and surface. Each repetition synthetic cycle leads to the addition of one more layer of monomers in the branches, called *generation*. Therefore, the generation number of the dendrimer is equal to the number of repetition cycles performed and to the number of branching points present from the core towards the periphery.

The first example of an iterative synthetic approach toward dendrimers has been reported in 1978 by Vögtle [3], who called it "cascade synthesis". In the mid-1980s, Tomalia [4] and Newkome [5] independently reported the divergent synthesis of two new families of dendrimers: poly(amidoamine) and the so-called "arborols", respectively. A growing interest on these macromolecules lead in 1990 to the convergent synthesis of aromatic polyether dendrimers by Fréchet [6]. The two different synthetic approaches can be explained as follows:

(a) In the divergent method, dendrimers are built from the core out to the periphery and in each step a new layer of branching units is added.
(b) The convergent method follows the opposite path: the skeleton of a dendron, defined as an entire branch, is built up step by step and finally reacted with the core moiety.

Dendrimer chemistry is nowadays a rapidly expanding field, as testified by the exponentially increasing number of papers published on this topic per year. Dendrimers keep attracting the attention of the scientific community because of their fascinating structure and unique properties: indeed the field of dendrimer

M. Baroncini, *Design, Synthesis and Characterization of new Supramolecular Architectures*, Springer Theses, DOI: 10.1007/978-3-642-19285-2_2,
© Springer-Verlag Berlin Heidelberg 2011

chemistry has evolved from the initial pursue of synthesizing new large and aesthetically pleasant molecules, to their characterization and finally it has now moved towards functionality. Today dendrimers are used or are planned to be exploited in a variety of applications, taking advantage of the great number of functional units that can be incorporated inside them, their tree-like structure containing internal dynamic cavities, their well-defined dimensions close to that of important biological molecules (Fig. 1.1), like proteins and bioassemblies, the presence of an internal microenvironment different from the bulk of the solution, and their endo- and exo-receptor properties. As a result, applications ranging from the biological and medical field (artificial enzymes, drug-delivery and diagnostics systems) to nanoengineering (molecular wires, light-emitting diodes), optical data transport (fiber optics), catalysis, energy-harvesting devices and mimics of natural photosynthesis are foreseen (for some recent reviews, see: [7–13]).

2.2 Dendrimers and Light

Currently, dendrimer research is developing swiftly in the direction of highly functional materials. Also in the field of photoactive dendrimers the complexity of the systems has increased enormously. The investigation of dendritic structures functionalized with luminescent groups [14], photoswitchable units [15], energy and/or electron donor–acceptor components and the implementation of such functionalized dendrimers in devices [16], provide insight in the fundamental processes occurring in such complex systems and in their future applications.

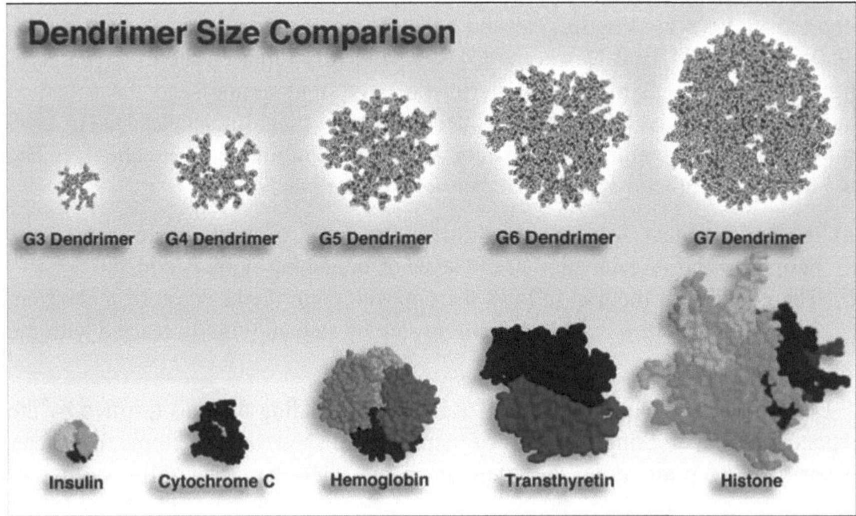

Fig. 1.1 Dendrimer size in comparison with important biological molecules

Since dendrimers can be functionalized with multiple chromophoric groups, that are in very close proximity, novel properties can arise compared to the single chromophoric system. Due to the stepwise synthesis, either divergent or convergent, chromophores can be implemented in the dendritic structure with high precision. The number of chromophores and the size of the dendrimer are very well controlled, which is of very great importance for some biological and biomedical applications. The presence of multiple chromophores in the same molecule enables the detection of single dendrimer via single molecule spectroscopy (SMS), which is particulary interesting for nanotechnology. Furthermore, an increased sensitivity with respect to specific classes of molecules can be established, enabling the detection of very low concentrations of these molecules. This is of great interest for the field of molecular recognition, e.g. for the development of biosensors (immuno-diagnostic).

A specific advantage of the dendritic framework is that a microenvironment can be crated around a single chromophore. By placing such protective environment around a chromophore its luminescence can be dramatically improved. Dendritic substituents can also promote supramolecular organization of chromophores, e.g. inducing the formation of fibers or doughnut-like structures.

The possibility to functionalize chromophores with large dendritic substituents is particularly interesting for the development of light emitting diodes (LEDs). The dendritic wedges do not only prevent the aggregation of chromophores, thereby reducing the amount of self quenching, but they also provide a way to improve the solubility of the chromophores in polymers, rendering a more homogeneous blend.

The introduction of photoisomerizable groups, such as azobenzene derivatives, in dendrimers enables the controlled induction of a structural change, especially when those units are attached to the core or implemented in the branches. If attached at the periphery, these photoisomerizable groups, can be used to "close" the surface of a dendrimer by means of a photoinduced increase of steric hindrance at the periphery. This type of dendrimers can be used as carriers of small molecules, while a controlled release of those guest molecules is possible using light, which induced the isomerization from *cis* to *trans*. In addition, azobenzene-functionalized materials are widely used in the field of datastorage.

The implementation of chromophores in dendritic structures can also provide more insight in the structural features of dendrimers. Dyes can be used as internal probes to investigate the microenvironment created by dendritic branches. At the same time the influence of external factor, e.g. the solvent or ions, on the microenvironment can be studied. This may concern a change in the conformation of the dendritic structure, but also the accessibility of the dendritic structure by other molecules. The implementation of multiple chromophores within one dendrimers allows the investigation of internal interactions between the chromophores, such as the formation of excimers and energy and electron transfer processes. To what extend these interactions take place will depend on the flexibility of the dendritic framework and the position of the chromophores within the dendritic structure.

2.3 Molecular Machines

What would be the utility of such machines? Who knows? I cannot see exactly what would happen, but I can hardly doubt that when we have some control of the arrangement of things on a molecular scale we will get an enormously greater range of possible properties that substances can have, and of the different things we can do.[17, 18].

2.3.1 The Concept of a Machine at the Molecular Level

In everyday life we make extensive use of (macroscopic) machines. A machine is [19] an apparatus for applying mechanical power, having several parts each with a definite function. When a machine is working, at least some of its components display changes in their relative positions. A machine is characterized by (1) the kind of energy input supplied to make it work, (2) the type of movements performed by its components, (3) the manner in which its operation can be monitored and controlled, (4) the possibility to repeat the operation at will and establish a cyclic process, (5) the timescale needed to complete a cycle of operation, and (6) the function performed by the machine. The concept of a machine can be extended to the molecular level. A molecular-level machine can be defined as an assembly of a discrete number of molecular components designed to perform mechanical-like movements (output) as a consequence of appropriate external stimuli (input). Although there are many chemical compounds whose constitutions and/or shapes can be modified by an external stimulus—for example, molecules capable of undergoing photoinduced cis/ trans isomerizations of their C=C, C=N, or N=N bonds—the term molecular-level machine will only be used for systems whose component parts undergo movement with relatively large amplitudes. Furthermore, systems in which the molecular movements are not controlled by some easily identifiable and well-characterized external stimulus will not be considered to be molecular-level machines. The extension of the concept of a machine to the molecular level is of interest not only for the sake of basic research, but also for the growth of nanoscience and the subsequent development of nanotechnology. The concept of a machine at the molecular level is not a new one. Our bodies can be viewed as very complex ensembles of molecular-level machines that power our physical motions in a multitude of different guises, repair tissue damage in a wide spectrum of situations and circumstances, as well as preside over our innermost worlds where we are preoccupied by our sensory perceptions, emotional states, and thought processes. The idea of constructing artificial molecular-level machines, however, is quite a recent one. The first time the topic was seriously contemplated was in 1959 by Richard Feynman in his historic address There is Plenty of Room at the Bottom to the American Physical Society in December of that year. The earliest examples of synthetic molecular-level machines, based on the photoisomerization of azobenzene, were reported [20] in the early 1980s. In the last 15 years research in the field of artificial molecular-level machines has been stimulated by several major scientific breakthroughs and paradigm shifts: they include (1) the rapid development of probe microscopies [21, 22] following the award of the Nobel Prize in Physics to Binnig and Rohrer in 1986; (2) a growing interest in supramolecular chemistry [23–28] after the award of the 1987 Nobel Prize in Chemistry to Pedersen, Cram, and Lehn; (3) the elucidation and unraveling of the working mechanisms of some key biological devices and machines [29–38], such as those involved in photosynthesis (Deisenhofer, Huber, and Michel recognized by the 1988 Nobel Prize in Chemistry) and in the ATP synthesis (leading to the 1997 Nobel Prize in Chemistry to Boyer, Skou, and Walker); (4) the great progress in understanding the mechanisms of the homogeneous and heterogeneous thermal and photoinduced electron-transfer reactions [39, 40] provided by Marcus who was awarded the Nobel Prize in Chemistry in 1992; and (5) the realization that the (physical) top-down approach to miniaturization in the electronics industry, for example, has intrinsic limitations and the increasing confidence that it can be replaced profitably by a (chemical) bottom-up approach [41]. In the past few years interest in artificial molecular machinery has grown exponentially and several short reviews covering specific aspects of the field are now available [42–44]. The aim of this article is to present a unified view of the field by focusing in on past achievements, present limitations, and future perspectives.

2.3.2 Defining Molecular-Level Machines

We have already identified the features and characteristics of macroscopic machines in, and have also established that one of the operational requirements of a molecular machine will be that the movement of its component parts will have relatively large amplitudes—a property which implies the occurrence of chemical reactions. In his 1959 address to the American Physical Society, Feynman noted that an internal combustion engine of molecular size is impossible. Other chemical reactions, liberating energy when cold, can be used instead. Even relatively cold chemical reactions, however, can destroy the molecules constituting a machine. Since such a machine works by repeating cycles (Point 4), an important requirement is that any chemical change or reaction taking place in the system has to be reversible. Within this constraint, any kind of chemical process that causes motions of a machine's component parts—for example, isomerizations, acid/base reactions, oxidation/reduction processes, complexation/decomplexation equilibria involving, for example, the making and breaking of hydrogen bonds—can be useful. Most chemical reactions occur as a result of thermal activation on mixing the reactants. If a molecular-level machine has to work by thermal activation it will need the addition of reagents at all steps in its working cycle, since the added reagents play the role (Point 1) of chemical energy inputs. Although such kinds of input can be useful, clearly the repeated addition of reagents will result in the accumulation of by-products that, after a relatively small number of cycles, will compromise the operation of the machine unless the products can be removed from the system, which is not an easy task to perform. In principle, the best energy inputs to make a molecular machine work (Point 1) are photons and electrons (or holes). Indeed, with appropriately chosen photochemically or electrochemically driven reactions, it is possible to design interesting and intriguing molecular machines. The motions performed by the component parts of a molecular-level machine (Point 2) depend to some extent on whether the machine is molecular or supramolecular in nature (the most authoritative and widely accepted definition (see Sect. 1.1 and [45, 46]) of supramolecular chemistry is the chemistry beyond the molecule, bearing on the organized entities of higher complexity that result from the association of two or more chemical species held together by intermolecular forces. On the basis of this definition, it is tempting to classify molecular machines as either molecular or supramolecular. However, it is not as simple and straightforward as one might think and so we avoid this classification. The fundamental nature of all the systems we discuss is their multicomponent nature. For any kind of device—be it macroscopic or otherwise—the observed function results from the cooperative interactions between the various component parts). Movements of component parts within classical molecules will necessarily involve changes in their conformations and/or configurations around covalent bonds that are formally single or double, respectively, in their orders, although these changes in molecular structure are accompanied by the making and breaking of intramolecular noncovalent bonds. Movements of the component parts within supermolecules (complexes) may be accompanied by conformational and/or configurational changes within their covalently linked molecular components; however, it will be largely the reorganization of intermolecular noncovalent bonds between the molecules that will usually reflect and constitute the movements within these kinds of molecular-level machines. In nonclassical so-called interlocked molecules the mechanical bonds that link the component parts together offer the close to unique opportunity, within relatively small molecules at least, to effect movements with large amplitudes upon their components, mainly as a result of the making and breaking of intercomponent noncovalent bonds. In order to monitor and control the operation (Point 3) of a molecular machine the motions of the component parts should bring about readable changes in some properties of the system. Any kind of chemical or physical probe can be useful in providing read-outs, particularly the various different types of spectroscopies that are currently available to us. In this regard, it should be pointed out that photochemistry and electrochemistry are often very useful since both photons and electrons can play the dual role of writing, that is, causing the change in the system, and reading, that is, reporting the resulting state of the system. The operating timescale (Point 5) of molecular machines is that of nuclear motions, which can range from nanoseconds to seconds depending on the nature of the components involved and the type of motions that are happening. We would like to point out that the description of a motion implies the definition of a fixed reference system. In the case of molecular machines, this matter is not so trivial. Finally, the functions that can be performed by exploiting the movements of the components in artificial machines (Point 6) are largely unpredictable.

2.4 Molecular Logic

The realization of computers with very small size, low power consumption, and computing performance that may be hard to reach with silicon-based technology ([47], see also [48]) is a strong motivation for the search for information-processing strategies based on molecules. The rational basis for this research stems from the fact that in living organisms information is transported, elaborated, and stored by molecular or ionic substrates [49, 50]. Although the components of a molecular computer will not necessarily have to operate analogously to microelectronic circuits [51], several efforts have been devoted to the design, synthesis, and characterization of chemical systems that mimic the operation of semiconductor logic gates [52]. As molecular switches convert input stimulations into output signals [53], the principles of binary (Boolean) logic [54] can be applied to the signal transduction operated by molecules under appropriate conditions (a large number of chemical switches exhibiting interesting properties, from a binary-logic viewpoint, have been reported in the literature. In many cases, the authors were either not aware of such behavior in their systems or published the work before the notion of molecular logic gates became popular. For the first explicit description of the analogy between molecular switches and logic gates, see: [55]). Implementation of the most common Boolean functions (PASS, YES, NOT, AND, NAND, OR, NOR, XOR, XNOR, and INH) with chemical systems is now possible. A critical issue of molecular logic gates is the interconnection of basic elements to create complex circuits. In contrast, electronic logic gates can be easily interconnected, owing to full input/output homogeneity. However, rather than relying on extensive physical connection of elementary gates, the construction of molecular logic networks can take advantage of functional integration and reconfiguration within a single molecule, which can be achieved by rational chemical design [56]. The fact that a relatively simple, commercially available dye molecule in aqueous solution can perform both the full-adder and full-subtractor functions—which in silicon-based systems require circuits made of five interconnected gates—is a demonstration of this idea. Recent reports on the first molecular versions of a digital multiplexer [57] and of a keypad access device have taken molecular logic one step further.

The development of novel computational architectures constitutes the main scientific driving force for the imitation of Boolean logic functions with molecular systems. Aside from futuristic speculations related to the construction of a chemical computer, recent work has shown that molecular logic gates could lead to practical applications in the not-too-distant future. Molecular devices that control protein folding [58] or release a chemical species [59] by processing chemical inputs according to programmed logic functions have been reported. They can be regarded as precursors of systems operating in vivo that are capable of autonomously diagnosing a disease and effecting a therapy. Moreover, a method based on molecular logic gates for tagging and identifying small objects in a large population has been proposed as a ready-to-use application for combinatorial chemistry [60, 61]. It should be noted that computing devices based on (supra)molecular species and soft matter represent a radically different approach to information processing with respect to computers made from solid-state semiconductors. Therefore, comparisons between these types of systems should be made with care and, for certain aspects, may not make much sense. It seems fair to state, however, that the examples mentioned in the previous paragraph are simple computational tasks that molecules can do and silicon cannot. Certainly, the investigation of intelligent molecules capable of elaborating signals introduces new concepts in the field of chemistry and stimulates research in the bottom-up approach to nanodevices.

References

 1. Fréchet JMJ, Tomalia DA (eds) (2001) Dendrimers and other dendritic polymers. Wiley, Chichester
 2. Newkome GR, Moorefield C, Vögtle F (2001) Dendrimers and dendrons: concepts, syntheses, perspectives. VCH, Weinheim
 3. Buhleier EW, Wehner W, Vögtle F (1978) Synthesis 2:155
 4. Tomalia DA, Baker H, Dewald JR, Hall M, Kallos G, Martin S, Roeck J, Ryder J, Smith P (1985) Polym J 17:117
 5. Newkome GR, Yao Z-Q, Baker GR, Gupta K (1985) J Org Chem 50:2003
 6. Hawker CJ, Fréchet JMJ (1990) J Am Chem Soc 112:7638
 7. Mery D, Astruc D (2006) Coord Chem Rev 250:1965
 8. Tomalia DA, Fréchet JMJ (eds) (2005) Special issue: dendrimers and dendritic polymers. In: Prog Polym Sci 30 (3–4)
 9. Scott RWJ, Wilson OM, Crooks RM (2005) J Phys Chem B 109:692
10. Chase PA, Klein Gebbink RJM, van Koten GJ (2004) Organomet Chem 689:4016
11. Ong W, Gomez-Kaifer M, Kaifer AE (2004) Chem Commun 1677
12. Ballauff M, Likos CN (2004) Angew Chem Int Ed 43:2998
13. Caminade A-M, Majoral J-P (2004) Acc Chem Res 37:341
14. Ceroni P, Bergamini G, Marchioni F, Balzani V (2005) Prog Polym Sci 30:453 and reference therein
15. Liao L-X, Stellacci F, McGrath DV (2004) J Am Chem Soc 126:2181
16. Thomas KRJ, Thompson AL, Sivakumar AV, Bardeen CJ, Thayumanavan S (2005) J Am Chem Soc 127:373
17. Feynman RP (1960) Eng Sci 23:22–36
18. Feynman RP (1960) Saturday Rev 43:45–47
19. Hawkins JM (1979) The Oxford paperback dictionary. Oxford University Press, Oxford
20. Shinkai S, Manabe O (1984) Top Curr Chem 121:76–104
21. Binnig G, Rohrer H (1987) Angew Chem 99:622–631
22. Binnig G, Rohrer H (1987) Angew Chem Int Ed Engl 26:606–614
23. Lehn J-M (1988) Angew Chem 100:91–116
24. Lehn J-M (1988) Angew Chem Int Ed Engl 27:89–112
25. Cram DJ (1988) Angew Chem 100:1041–1052

26. Cram DJ (1988) Angew Chem Int Ed Engl 27:1009–1020
27. Pedersen CJ (1988) Angew Chem 100:1053–1059
28. Pedersen CJ (1988) Angew Chem Int Ed Engl 27:1021–1027
29. Deisenhofer J, Michel H (1989) Angew Chem 101:872–892
30. Deisenhofer J, Michel H (1989) Angew Chem Int Ed Engl 28:829–847
31. Huber R (1989) Angew Chem 101:849–871
32. Huber R (1989) Angew Chem Int Ed Engl 28:848–869
33. Boyer PD (1998) Angew Chem 110:2424–2436
34. Boyer PD (1998) Angew Chem Int Ed 37:2296–2307
35. Walker JE (1998) Angew Chem 110:2438–2450
36. Walker JE (1998) Angew Chem Int Ed 37:2308–2319
37. Skou JC (1998) Angew Chem 110:2452–2461
38. Skou JC (1998) Angew Chem Int Ed 37:2320–2328
39. Marcus RA (1993) Angew Chem 105:1161–1172
40. Marcus RA (1993) Angew Chem Int Ed Engl 32:1111–1121
41. Nalwa HS (ed) (2000) Handbook of nanostructured materials and nanotechnology. Academic Press, New York
42. Balzani V, Credi A, Venturi M (1999) In: Ungaro R, Dalcanale E (eds) Supramolecular science: where it is and where it is going. Kluwer Academic, Dordrecht, pp 1–22
43. Balzani V, Credi A, Venturi M (2000) In: Shibasaki M, Stoddart JF, Vögtle F (eds) Stimulating concepts in chemistry. Wiley-VCH, Weinheim, pp 255–266
44. Balzani V, Gómez-López M, Stoddart JF (1998) Acc Chem Res 31:405–414
45. Lehn J-M (1990) Angew Chem 102:1347–1362
46. Lehn J-M (1990) Angew Chem Int Ed Engl 29:1304–1319
47. International Technology Roadmap for Semiconductors (ITRS) 2005 Edition and 2006 Update. http://www.itrs.net
48. Thompson SE, Parthasarathy S (2006) Mater Today 9:20
49. Goodsell DS (2004) Bionanotechnology: lessons from nature. Wiley, Hoboken
50. Jones RAL (2004) Soft machines: nanotechnology and life. Oxford University Press, New York
51. Tullo A (2006) Chem Eng News 84:22
52. Balzani V, Credi A, Venturi M (2003) Molecular devices and machines: a journey into the nanoworld. Wiley-VCH, Weinheim, Chap 9
53. Feringa BL (ed) (2001) Molecular switches. Wiley-VCH, Weinheim
54. Gregg JR (1998) Ones and zeros: understanding boolean algebra, digital circuits, and the logic of sets. Wiley, New York
55. de Silva AP, Gunaratne HQN, McCoy CP (1993) Nature 364:42
56. de Silva AP (2005) Nat Mater 4:15
57. Andréasson J, Straight SD, Bandyopadhyay S, Mitchell RH, Moore TA, Moore AL, Gust D (2007) Angew Chem 119:976
58. Muramatsu S, Kinbara K, Taguchi H, Ishii N, Aida T (2006) J Am Chem Soc 128:3764
59. Amir RJ, Popkov M, Lerner RA, Barbas III CF, Shabat D (2005) Angew Chem 117:4452
60. de Prasanna Silva A, James MR, McKinney BOF, Pears DA, Weir SM (2006) Nat Mater 5:787
61. Webb R (2006) Nature 443:39

Chapter 3
Materials and Methods

3.1 Materials

All chemicals were purchased from Aldrich and were used without further purification. Solvents were purchased from Aldrich and purified according to literature procedures. Thin-layer chromatography (TLC) was performed on aluminum sheets coated with silica gel 60F (Merck 5554). The plates were inspected by UV light. Column chromatography was performed on silica gel 60 (Merck 40–60 nm, 230–400 mesh). High-performance liquid chromatography (HPLC) grade MeCN was purchased from Aldrich and degassed by bubbling He. Known amounts of the compounds were dissolved in MeCN (HPLC grade) to afford stock solution of known concentration (0.001 M). The resulting solutions (injected volume 10 µL) were analyzed at ambient temperature by HPLC (flow rate 1.0 mL min^{-1}; mobile phase, pump A 0.1% CF_3CO_2H in H_2O, pump B MeCN/0.1% CF_3CO_2H in H_2O/95:5; time (min)/pump A (%) = 0/100, 8/100, 28/0, 42/0, 45/100) by employing a Hypersil BDS C18 column (length 25 cm, inside diameter 5.6 cm) operated by HP 1090 (Hewlett–Packard) connected to thermo surveyor UV–Vis diode array detector. Melting points were determined on an Electrothermal 9200 apparatus and reported uncorrected. ^{1}H and ^{13}C spectra were recorded on Varian Mercury 400 (400 and 100 MHz, respectively), using residual solvents as the internal standard. Samples were prepared using deuterated solvents purchased from Cambridge Isotope Laboratories. The chemical shift values were expressed as δ values, and the couplings constants (J) are in Hertz. The following abbreviations were used for signal multiplicities: s, singlet; d doublet; t triplet; m multiple; br broad. Electron impact mass spectra (EIMS) were obtained from VG Prospec mass spectrometer.

The compounds described in Chap. 4 were synthesized according to the following procedures:

5.4PF$_6$ (**B2**.4PF$_6$): A solution of **4**.2PF$_6$, prepared following literature procedure [1], (1.0 g, 1.38 mmol) in dry MeCN (10 mL) was added dropwise to a solution of

M. Baroncini, *Design, Synthesis and Characterization of new Supramolecular Architectures*, Springer Theses, DOI: 10.1007/978-3-642-19285-2_3,
© Springer-Verlag Berlin Heidelberg 2011

3 [2] (1.0 g, 3.45 mmol) in dry MeCN (10 mL) under reflux and an argon atmosphere during 2 h. The reaction mixture was stirred for 2 days and then cooled to ambient temperature, the solvent was removed under vacuum and the residue subjected to column chromatography (SiO_2, MeOH:2M NH_4Cl_{aq}:$MeNO_2$, 7:2:1). The excess of $MeNO_2$ and MeOH were then removed under vacuum such that the product still remained dissolved and a saturated solution of NH_4PF_6 was added dropwise. The solid was filtered off, washed with H_2O (4 × 50 mL) and dried to afford **5**.$4PF_6$ as a yellow solid (1.3 g, 66.3% over two steps). Melting point > 250°C. ^1H NMR (500 MHz, $(CD_3)_2CO$) δ = 9.45 (d, J = 10 Hz, 4H), 9.41 (d, J = 10 Hz, 4H), 8.76 (d, J = 10 Hz, 8H), 7.93 (s, 1H), 7.89 (s, 2H), 7.66 (d, J = 8.5 Hz, 4H), 7.08 (d, J = 6.5 Hz, 4H), 6.21 (s, 4H), 6.11 (s, 4H), 4.74 (s, 2H), 4.21–4.17 (m, 4H), 3.86–3.82 (m, 4H), 3.68–3.65 (m, 4H), 3.52–3.50 (m, 4H), 3.31 (s, 6H). HR-ESI: m/z = 1299.32 $[M - PF_6]^+$.

6.$4PF_6$: The alcohol **5**.$4PF_6$ (2 g, 1.38 mmol) was dissolved in dry THF:MeCN (3 mL:1 mL) and CBr_4 (2.3 g, 6.9 mmol) and Ph_3P (1.82 g, 6.9 mmol) were added under vigorous stirring. The reaction mixture was stirred under argon at room temperature to completion, during which 1.25 equiv amounts of CBr_4 and Ph_3P at 30 min intervals were added until TLC showed no starting material. The reaction mixture was then poured into H_2O, extracted with CH_2Cl_2 (3 × 25 mL) and the combined extracts were evaporated to dryness. The residue was subjected to column chromatography (SiO_2, Me_2CO, and then 5 mM solution of NH_4PF_6 in Me_2CO). The resulting product was washed with H_2O and dried to afford the desired product (1.27 g, 61%) as a dark brown solid. Melting point > 250°C. ^1H NMR (500 MHz, $(CD_3)_2CO$) δ = 9.42 (d, J = 10 Hz, 4H), 9.37 (d, J = 9 Hz, 4H), 8.72 (d, J = 6 Hz, 8H), 7.86 (s, 1H), 7.84 (s, 2H), 7.61 (d, J = 10 Hz, 4H), 7.04 (d, J = 6 Hz, 4H), 6.16 (s, 4H), 6.15 (s, 4H), 4.58 (s, 2H), 4.14–4.13 (m, 4H), 3.79–3.77 (m, 4H), 3.61–3.60 (m, 4H), 3.47–3.45 (m, 4H), 3.25 (s, 6H). HR-ESI: m/z = 1363.26 $[M - PF_6]^+$.

8.$18PF_6$ (**B9**.$18PF_6$): A solution of **7**.$3PF_6$ [3] (73.5 mg, 0.072 mmol) in dry MeCN (1 ml) was added dropwise to a solution of **6**.$4PF_6$ (647.3 mg, 0.43 mmol) in dry MeCN (2 mL) under reflux and an argon atmosphere during 2 h. The reaction mixture was stirred for 48 h and then cooled to ambient temperature, the solvent was removed under vacuum and the residue subjected to HPLC (Hypersil BDS C18 column, flow rate 1.0 mL min^{-1}; mobile phase, pump A 0.1% CF_3CO_2H in H_2O, pump B MeCN/0.1% CF_3CO_2H in H_2O/95:5; time (min)/pump A (%) = 0/100, 8/100, 28/0, 42/0, 45/100). The excess of MeCN and H_2O was evaporated off under vacuum such that the product still remained dissolved, and saturated aqueous solution of NH_4PF_6 was added dropwise. The solid was filtered off, washed with H_2O (4 × 20 mL) and dried under vacuum to afford **8**.$18PF_6$ as a light brown solid (616 mg, 25%). Melting point > 250°C. ^1H NMR (500 MHz, $(CD_3)_2CO$) δ = 8.91–8.87 (m, 36H), 8.38–8.31 (m, 36H), 7.62 (br s, 9H), 7.60 (br s, 3H), 7.44 (d, J = 8.5 Hz, 12H), 7.00 (d, J = 7.5 Hz, 12H), 5.80–5.72 (m, 36H), 4.11–4.09 (m, 12H), 3.76–3.74 (m, 12H), 3.60–3.58 (m, 12H), 3.47–3.45 (m, 12H), 3.26 (s, 18H). ^{13}C NMR (125 MHz, CD_3CN) δ = 160.1, 150.5, 150.4, 149.9, 145.7, 145.6, 145.2, 134.7,

134.6, 131.7, 131.2, 127.3, 127.2, 124.2, 115.3, 71.4, 70.0, 68.9, 67.6, 63.5, 57.8. HR-ESI: m/z = 1289.80 $[M - 4PF_6]^+$.

9.12PF$_6$: A solution of **4**.2PF$_6^{[1]}$ (0.20 g, 0.27 mmol) in dry MeCN (2 mL) was added dropwise to a solution of **6**.4PF$_6^{[2]}$ (1.44 g, 0.95 mmol) in dry MeCN (2 mL) under reflux and an argon atmosphere during 2 h. The mixture was stirred for 48 h and then cooled to ambient temperature, the solvent was removed under vacuum and the residue subjected to column chromatography (SiO$_2$, Me$_2$CO and then 5, 10 mM and saturated solutions of NH$_4$PF$_6$ in Me$_2$CO, consecutively). The resulting product was washed H$_2$O and dried to afford the alcohol **9**.12PF$_6$ (2.47 g, 65%). Melting point > 250°C. ^{1}H NMR (500 MHz, (CD$_3$)$_2$CO) δ = 9.05–8.9 (m, 24H), 8.5–8.3 (m, 24H), 7.75–7.60 (br s, 9H), 7.5 (d, J = 6 Hz, 8H), 7.05 (d, J = 6 Hz, 8H), 5.9–5.7 (m, 24H), 4.7 (s, 2H), 4.25–4.10 (m, 8H), 3.85–3.70 (m, 8H), 3.65–3.60 (m, 8H), 3.55–3.45 (m, 8H), 3.30 (s, 12H). HR-ESI: m/z = 1843.2 $[M - 2PF_6]^+$, 1180.1 $[M - 3PF_6]^+$, 849.5 $[M - 4PF_6]^+$.

11.42PF$_6$ (**B21**.42PF$_6$): The alcohol **9**.12PF$_6$ (500 mg, 0.126 mmol) was dissolved in dry THF:MeCN (3 mL:1 mL) and CBr$_4$ (62 mg, 0.188 mmol) and Ph$_3$P (49 mg, 0.188 mmol) were added under vigorous stirring during 30 min. The reaction mixture was stirred under argon and room temperature to completion, during which 1.25 equiv amounts of CBr$_4$ and Ph$_3$P at 30 min intervals were added until TLC showed no starting material. The reaction mixture was poured into H$_2$O, extracted with CH$_2$Cl$_2$ (3 × 25 mL) and the combined extracts were dried and evaporated to dryness. The residue subjected to column chromatography (SiO$_2$, Me$_2$CO and then 5, 10 mM and saturated solutions of NH$_4$PF$_6$ in Me$_2$CO, consecutively). The resulting product was washed with H$_2$O and dried to afford the desired product (203 mg, 40%) as a dark brown solid, which was added to a solution of **7**.3PF$_6^{[20]}$ (9 mg, 0.0084 mmol) in MeCN (1 mL). The reaction mixture was stirred for 5 days under reflux and an argon atmosphere then cooled to ambient temperature. The solvent was removed under vacuum and the residue subjected to HPLC Hypersil BDS C18 column, flow rate 1.0 mL.min; mobile phase, pump A 0.1% CF$_3$CO$_2$H in H$_2$O, pump B MeCN/ 0.1% CF$_3$CO$_2$H in H$_2$O/95:5; time (min)/pump A (%) = 0/100, 8/100, 28/0, 42/ 0, 45/100). The excess of MeCN and H$_2$O was evaporated off under vacuum such that the product still remained dissolved, and saturated aqueous solution of NH$_4$PF$_6$ was added dropwise. The solid was filtered off, washed with H$_2$O (4 × 20 mL) and dried under vacuum to afford **11**.42PF$_6$ as a light brown solid (0.1, 15%). Melting point > 250°C. ^{1}H NMR (500 MHz, (CD$_3$)$_2$CO) δ = 8.91–8.67 (m, 84H), 8.36–8.11 (m, 84H), 7.60 (br s, 27H), 7.56 (br s, 3H), 7.42 (d, J = 8.5 Hz, 24H), 6.90 (d, J = 7.5 Hz, 24H), 5.78–5.70 (m, 84H), 4.00–3.69 (m, 24H), 3.70–3.68 (m, 24H), 3.56–3.45 (m, 24H), 3.42–3.37 (m, 24H), 3.24 (s, 36H). ^{13}C NMR (125 MHz, CD$_3$CN) δ = 160.9, 146.4, 146.0, 135.6, 135.5, 132.5, 132.0, 128.2, 128.0, 125.0, 118.0, 117.5, 116.1, 72.3, 70.8, 69.8, 68.4, 65.1, 64.3, 58.6.

The compounds described in Chap. 5 were synthesized by the group of Prof. Fritz Vogtle at University of Bonn.

The compounds described in Chap. 6 were synthesized by A. Arduini, A. Pochini and A. Secchi at University of Parma.

The compounds described in Chap. 7 were synthesized by A. Arduini, A. Pochini and A. Secchi at University of Parma and by F. Raymo at University of Miami.

The compounds described in Chap. 8 were synthesized according to the following procedures:

EE-1H·PF$_6$. A solution of 4-nitrosotoluene (3.63 g, 30.0 mmol) and bis(4-aminobenzyl)amine (3.41 g, 15 mmol) in acetic acid (50 mL) was stirred under argon at room temperature for 12 h in the dark. Water was added to the reaction mixture and the dark red precipitate formed was collected by filtration and washed with water. The product was recrystallized from boiling DMF, filtered and washed with ether. The isolated salt was dissolved in acetone and a saturated solution of NH_4PF_6 in water was added to precipitate the PF_6 salt as a brown-orange powder. The product was dried under vacuum to afford 6.3 g (72%) of *EE*-1H·PF$_6$. Melting point > 210°C decomposition. ^{1}H NMR (400 MHz, CD$_3$CN): = 7.95 (d, $J = 8.3$ Hz, 4H), 7.84 (d, $J = 8.4$ Hz, 4H), 7.67 (d, $J = 8.4$ Hz, 4H), 7.41 (d, $J = 8.3$ Hz, 4H), 4.36 (s, 4H), 2.44 (s 6H); ^{13}C NMR (100.5 MHz, CD$_3$CN): = 154.2, 151.6, 143.8, 133.8, 132.4, 131, 123.9, 123.8, 52.1, 21.5 ppm; MS (ESI): m/z (%): calcd 434.2345; found 434.2346 (100) $[M - PF_6]^+$.

The compounds described in Chap. 9 were synthesized according to the following procedures:

8-methoxyquinolinium hexafluorophosphate [8-MQ-H][PF$_6$]. 8-Hydroxy-quinoline (1.45 g. 10.0 mmol), KOH (2.53 g. 45.0 mmol), MeI (2.13 g. 15.0 mmol), and THF (100 mL) were stirred at room temperature for 12 h. After filtration and solvent removal, 8-methoxyquinoline (1.27 g, 80%) was isolated after flash chromatography on silica gel (hexane/EtOAc 3:1) as a red oil. The isolated oil was dissolved in acetone/water and the solution was acidified with 10% aqueous HCl. To the clear solution an excess of NH_4PF_6 was added in order to precipitate the hexafluorophosphate salt of 8-methoxy-quinolinium, that was collected by filtration. The salt was recrystallized from isopropyl alcohol to afford [8-MQ-H][PF$_6$] (2.39 g, 98%) as pale green needles. Melting point 141–142°C. ^{1}H NMR (400 MHz, Acetone-d$_6$): δ 9.35 (1H, d, $J = 5.3$ Hz), 9.25 (1H, d, $J = 8.4$ Hz), 8.24 (1H, dd, $J = 8.4$, 5.3 Hz), 7.95–7.96 (2H, m), 7.72 (1H, dd, $J = 5.8$, 3.0 Hz), 4.21 (3H, s); ^{13}C NMR (100.5 MHz, Acetone-d$_6$): δ 149.6, 148.0, 144.5, 131.2, 130.3, 129.5, 123.2, 120.5, 113.7, 56.9.

Preparation of the polystyrene films. A MeCN solution of 8-MQ was added to a CHCl$_3$ solution of polystyrene 1.5% (*w/w*), such that the concentration of the chromophore in the resulting mixture was 2.25×10^{-3} M. The solution was placed between two microscopy quartz coverslips (previously washed with ethanol and water, and carefully dried). After the solvent had evaporated, one coverslip was removed and the sample was subjected to fluorescence spectroscopic analysis. The thickness of the films was in the order of 0.7 μm.

3.2 Photophysical Techniques

3.2.1 Electronic Absorption Spectra

All the absorption spectra in the 190–1100 nm range were recorded at room temperature on solutions contained in quartz cuvettes (optical pathlength 1 and 5 cm, Hellma®) by using a Perkin Elmer λ 40 spectrophotometer. The precision on the wavelength values was ±2 nm. Molar absorption coefficient values were determined using the Lambert–Beer law; the experimental error, mostly due to weighting error, can be estimated to be around ±5%.

3.2.2 Luminescence Spectra

Fluorescence and phosphorescence emission and excitation spectra in the 250–900 nm range were recorded with Perkin Elmer LS 50 spectrofluorimeters equipped with Hamamatsu R928 or R955 photomultiplier. In case of weak luminescence signals the more sensitive Fluorolog 3 from ISA (Jobin-Yvon–Spex) was used.

Room temperature spectra were recorded in the same spectrofluorimetric suprasil quartz cuvettes described for the electronic absorption spectra. In order to have comparable luminescence intensity measurements some correction had to be applied to the experimental data. These corrections were introduced to take into account instrumental geometrical effects and the distribution of the exciting light among the species effectively present in the solution [4, 5].

Spectra in frozen matrix at 77 K were taken using quartz (or glass) tubes with an internal diameter of about 2 mm and a 20 cm length immersed in liquid nitrogen. A transparent dewar (glass or quartz) with a cylindrical terminal part with a 1 cm external diameter was employed. Such a device easily fit into the sample holder of the spectrofluorimeters above indicated. Luminescence spectra recorded in the 650–900 nm region were corrected for the non-linear response of the photomultiplier towards photons of different wavelength making reference to a previously experimentally determined calibration curve obtained. The precision on the wavelength values was ±2 nm.

Luminescence spectra in the near infrared (NIR) region were recorded by a home-made apparatus based on an Edinburgh CD900 spectrofluorimeter, which uses a Xenon lamp as the excitation source and a liquid nitrogen cooled hyperpure germanium crystal as a detector.

3.2.3 Luminescence Quantum Yield

Luminescence quantum yields were determined on solution samples at room temperature referring to the relative method optimized by Demas and Crosby. The quantum yield is expressed as:

$$\Phi_S = \Phi_R(A_S/A_R)(n_S/n_R)^2$$

where Φ, A and n indicate the luminescence quantum yield, the area subtended by the emission band (in the intensity versus frequency spectrum) and the refractive index of the solvent used for the preparation of the solution, respectively; the subscripts S and R stand for sample and reference, respectively. A_S and A_R must be relative to the same instrumental conditions ant to the same solution absorption at the excitation wavelength.

Different standards were selected depending on the spectral region of interests: naphthalene in degassed cyclohexane ($\Phi = 0.23$) [6], Fluorescein in NaOH 0.01 M ($\Phi = 0.90$) [7] or quinine sulphate in 1 M aqueous solution ($\Phi = 0.546$) [8]. The experimental error was $\pm 15\%$.

3.2.4 Luminescence Lifetime Measurements

Excited state lifetimes in the range from 0.5 ns to 30 µs were measured with an Edinburgh Instrument time correlated single-photon counting technique [9]. A schematic view of this instrument is reported in Fig. 3.1. The excitation impulse is obtained by a gas discharge lamp (model nF900, filled with nitrogen or deuterium, depending on excitation requirements) delivering pulses of 0.5 ns width at a frequency comprised between 1 and 100 kHz or by a pulsed diode laser (406 nm Picoquant). A photomultiplier tube (Hamamatsu R928P) cooled at $-20°C$ and suitably amplified is used as stop detector.

Fig. 3.1 Experimental set-up of the single photon counting

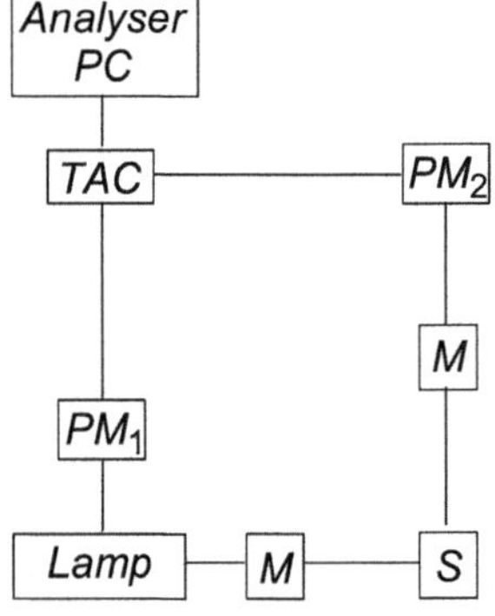

Lifetimes in the range between 10 µs and 5 s were measured with the same Perkin Elmer LS 50 spectrofluorimeter employed for the luminescence spectra acquisition. In this case the excitation pulse is generated by a Xe lamp (5–50 Hz) and the emission decay directly recorded. The experimental error on the lifetime measurements is $\pm 10\%$.

3.2.5 Titration Experiments

Titration experiments have been performed directly in the spectrofluorimetric cuvettes previously described. 2.5 ml of dendrimer solution was introduced in the cell with conventional pipettes and increasing amounts of a solution of the ion under examination added. Small volumes (of the order of 1–20 µl) were added in order to minimize the dilution effect which anyway was taken into account during data elaboration. Host concentrations were typically in the 10^{-5} to 10^{-6} M range while guest solutions, before dilution, were between 10^{-3} and 10^{-4} M. During titration, usually both absorption and luminescence spectra were recorded. When possible, the fluorophores were excited at wavelength where absorption changes during the titration were small in order to simplify the luminescence spectra corrections. For the additions Hamilton microlitre syringes were used.

3.2.6 Laser Flash Photolysis

Laser flash photolysis experiments were carried out with a home made apparatus that is schematised in Fig. 3.2. In our set-up a Surelite Nd:YAG laser (pulse width ≤ 10 ns) was used as excitation source and a hight-pressure Xenon arc-lamp was used as analysing light. The signal after the sample was captured by a monochromator/photomultiplier detection system, recorded by a Tektronix TDS640A digitizer oscilloscope and transferred to a PC computer.

Fig. 3.2 Laser flash photolysis experimental set-up

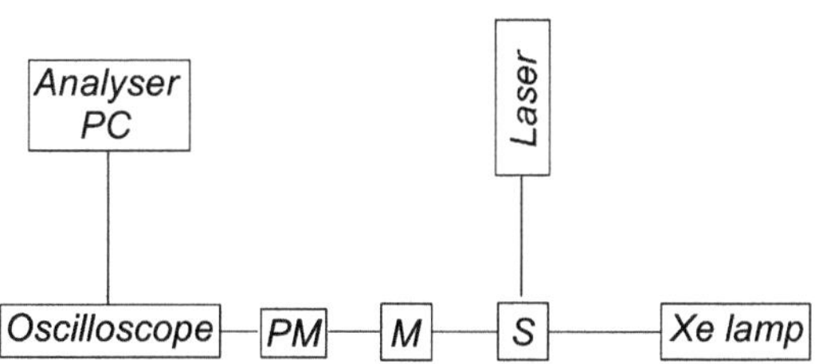

3.2.7 Photochemical Experiments

The experiments were carried out at room temperature on 3 ml solution (concentration in the range 10^{-5} to 10^{-4} M) contained in a quartz cell with optical pathlength of 1 cm. When necessary, the solution was degassed using the freeze–pump–thaw method (at least 3 cycles). The solution was continuously stirred during the irradiation with a Hellma Microver apparatus. Irradiation in the UV region was performed with a medium pressure Q400 Hanau mercury lamp (150 W); the wavelength of 287, 313, 365 and 436 nm were isolated by means of interference filters. The incident light intensity was measured using the ferric oxalate actinometer [10, 11]; and was of the order of 10^{-7} Nhν/min. Irradiation in the visible region was performed with a tungsten halogen lamp (150 W, 24 V).

Photoreaction quantum yield were determined by monitoring the changes in absorbance associated with the disappearance of the reactants or the formation of the products, at a wavelength where the absorbance could be related to the concentration of these species.

3.3 Electrochemical Experiments

3.3.1 Electrochemical Setup

Electrochemical experiments were carried out with an EcoChemie Autolab 30 multipurpose instrument interfaced to a personal computer.

The working electrode for the voltammetric experiments was a glassy carbon electrode (0.14 cm^2, Amel) or a Pt ultramicroelectrode ($r = 5$ μm); their surface was routinely polished with a 0.3 μm alumina–water slurry on a felt surface immediately prior to use.

The counter electrode was a Pt wire separated from the solution by means of a fine glass frit. This electrode was cleaned by burning on a Bunsen flame.

A silver wire were used in voltammetric studies as a reference and quasi-reference electrodes, respectively. In all cases a reference compound was added to the solution as a standard for potential values. The criteria for the choice of such a standard were (i) its chemical inertness towards the species under examination, and (ii) the ability to give reversible, well defined oxidation and/or reduction processes not overlapping with those of the sample. The most used standard was ferrocene, which is known to give a reversible monoelectronic oxidation process [12] and to be a relatively innocent species.

The solvent used was CH$_3$CN Romil of high-dry quality and taken under argon stream; the solution was continuously purged with argon during the experiments. The concentration of the examined compounds was around 5×10^{-4} M; 0.05 M tetraethylammonium hexa-fluorophosphate (Fluka puriss.) were added as supporting electrolyte. These salts were dried at 110°C for 3 days prior to use.

All the potential values reported in the following chapters are reduction potentials and referred to the SCE electrode, unless otherwise noted.

3.3.2 Cyclic Voltammetric Experiments

Cyclic voltammograms were obtained at sweep rates of 1–10 V s^{-1}. The criteria of (i) separation of less than 80 mV between cathodic and anodic peaks, (ii) close to unity ratio of the intensities of the cathodic and anodic currents, and (iii) constancy of the peak potential on changing sweep rate in the cyclic voltammograms were used to establish the reversibility of a process. For reversible processes, the halfwave potential value was calculated from the average of the potential values for anodic and cathodic peaks.

The number of electrons exchanged in a redox process has been evaluated by comparison of the current intensity of the corresponding voltammetric wave with that obtained for species undergoing redox processes which involve a known number of electrons. The following expression was used [13]:

$$n_S/n_R = (I_S C_R/I_R C_S)\,(M_S/M_R)^{0.275}$$

where n, I, C and M indicate the number of electrons exchanged in the process, the current intensity of the corresponding voltammetric wave, the concentration and the molecular mass, respectively; the subscripts S and R refer to sample and reference, respectively. The term containing the molecular masses apply the correction for differences in the diffusion coefficient of the electroactive species.

The experimental error on the potential values for reversible processes was estimated to be within ±5 mV. The diffusion coefficients were determined by chronoamperometric experiments using either a glassy carbon (0.14 cm^2, Amel) or a Pt ultramicroelectrode ($r = 5$ µm) as working electrode.

References

1. Wong EW, Collier CP, Behloradsky M, Raymo FM, Stoddart JF, Heath JR (2000) J Am Chem Soc 122:5831–5840
2. Collier CP, Jeppesen JO, Luo Y, Perkins J, Wong EW, Heath JR, Stoddart JF (2001) J Am Chem Soc 123:12632–12641
3. Amabilino DB, Ashton PR, Belohradsky M, Raymo FM, Stoddart JF (1995) J Chem Soc Chem Commun 751–753
4. Credi A, Prodi L (1996) EPA Newsl 58:50
5. Credi A, Prodi L (1998) Spectrochim Acta A 54:159
6. Berlman IB (1965) Handbook of fluorescence spectra of aromatic molecules. Academic, London
7. Fleming GR, Knight AWE, Morris JM, Morrison RJS, Robinson GW (1977) J Am Chem Soc 99:4306
8. Meech SR, Phillips D (1983) J Photochem 54:159

9. Lakowicz JR (1999) Principles of fluorescence spectroscopy, 2nd edn. Kluwer Academic, New York
10. Hatchard CG, Parker CA (1956) Proc R Soc Lond A 235:518
11. Fisher E (1984) EPA Newsl 21:33
12. Dubois D, Monitot G, Kutner W, Jones MT, Kadish KM (1992) J Phys Chem 96:7137
13. Flanagan JB, Margel S, Bard AJ, Anson FC (1978) J Am Chem Soc 100:4248

Chapter 4
Dendrimers: Polyviologen Dendrimers as Hosts and Charge-Storing Devices

4.1 Introduction

Dendrimers are repetitively branched, yet constitutionally well-defined macro-molecules, exhibiting high monodispersity [1–3]. These nanoscopic compounds can contain selected chemical units in predetermined sites of their structure (core, branches, periphery) whose mutual interactions can lead to unusual, sometimes unpredictable, chemico-physical properties resulting in a wide range of potential applications (for some recent reviews, see [4–11]). Dendrimers containing pho-toactive and/or electroactive moieties are currently attracting much attention since they can be designed to perform useful functions such as light-harvesting (for some recent reviews, see: [12–15]), charge-pooling [16–31], ion sensing with signal amplification [32–36], and molecular recognition (for some reviews and recent papers, see: [37–45]).

The dendrimers described here, $\mathbf{B9}^{18+}$ and $\mathbf{B21}^{42+}$ (Scheme 4.1), are based on a 1,3,5-trisubstituted benzenoide core, contain 9 and 21, respectively, 4,4′-bipyrid-inium dication (usually called viologen) units in their branches, and 6 and 12, respectively, aryloxy groups in the periphery. Viologen is a well known electro-active species that undergoes two successive, one-electron, reversible reduction processes [46–48] and exhibits peculiar spectroscopic features in both its dicat-ionic and radical-cationic forms. Viologen units are also known to give strong donor–acceptor complexes with electron donor species [48].

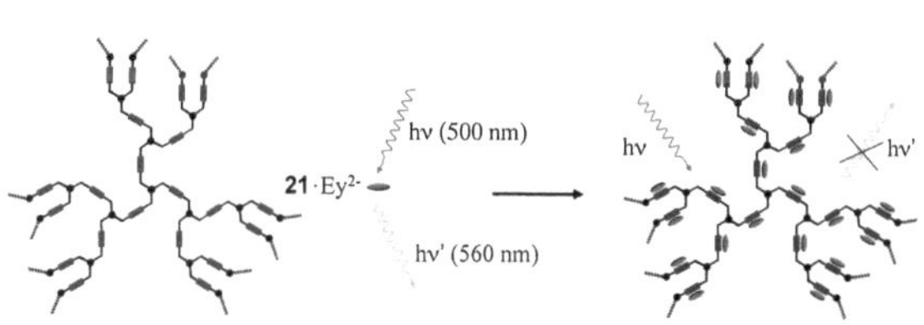

M. Baroncini, *Design, Synthesis and Characterization of new Supramolecular Architectures*, Springer Theses, DOI: 10.1007/978-3-642-19285-2_4,
© Springer-Verlag Berlin Heidelberg 2011

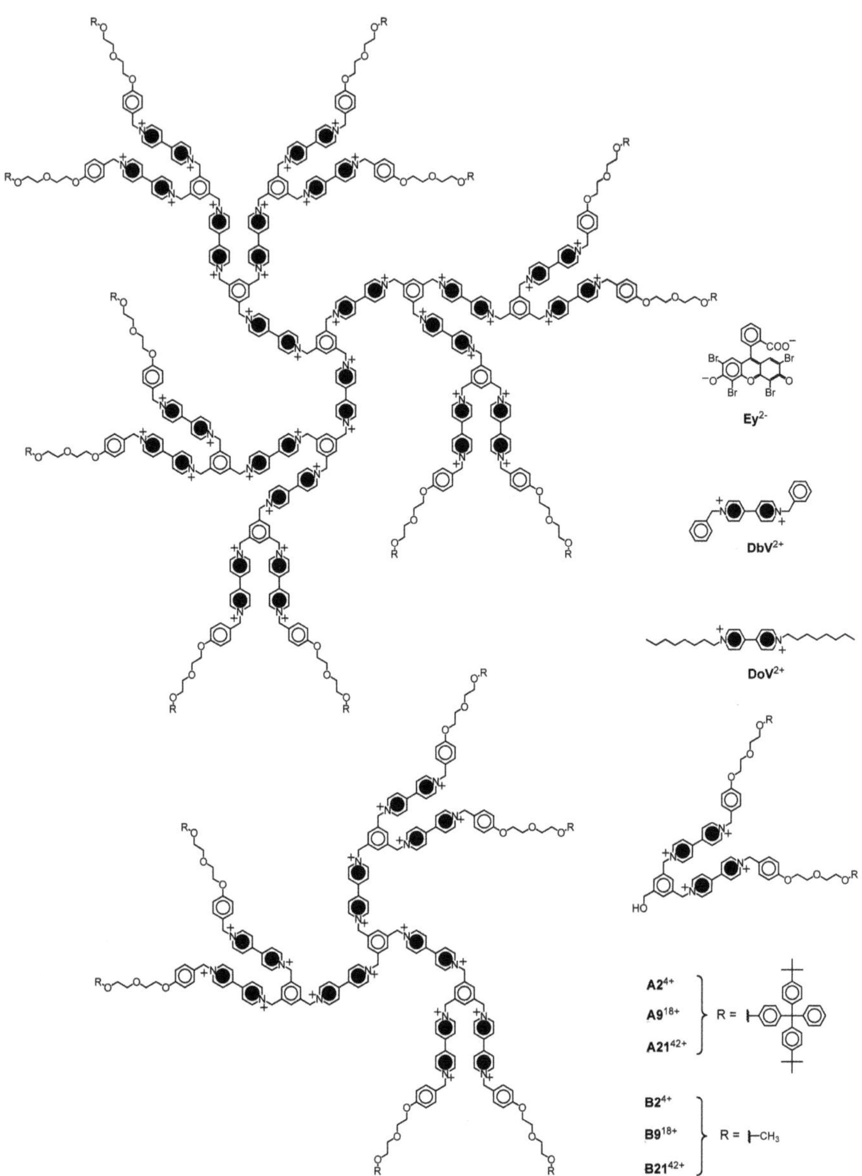

Scheme 4.1 Formulas of the examined compounds. The **B**n^{2n+} and **A**n^{2n+} symbols used indicate the number of the viologen units contained in the dendritic structure (**n**) and, as a superscript, the overall electric charge (2**n**) of each compound

In this chapter are described (i) the synthesis of dendrimers **B9**$^{18+}$ and **B21**$^{42+}$ as hexafluorophosphate salts (ii) their interaction with eosin dianion (**Ey**$^{2-}$, Scheme 4.1), and (iii) their electrochemical and photosensitized reduction. For comparison purposes, are also illustrated the properties of the 1,1′-dibenzyl-4,4′-

bipyridinium ($\mathbf{dbV^{2+}}$) and 1,1′-dioctyl-4,4′-bipyridinium ($\mathbf{doV^{2+}}$) mono-viologen compounds, and of dendron $\mathbf{B2^{4+}}$, which contains two viologen units (Scheme 4.1).

The results obtained are discussed in comparison with previously reported studies [18, 49] for two analogous dendrimers ($\mathbf{A9^{18+}}$ and $\mathbf{A21^{42+}}$ see Scheme 4.1) terminated with tetraarylmethane bulky moieties instead of aryloxy groups. The properties of dendron $\mathbf{A2^{4+}}$, analogous of $\mathbf{B2^{4+}}$, are also studied to complete the comparison between the two families of compounds.

4.2 Synthesis

The routes employed to synthesize the dendrimers $\mathbf{B9}.18PF_6$ and $\mathbf{B21}.42PF_6$, and their intermediates $\mathbf{6}.4PF_6$ and $\mathbf{10}.12PF_6$, are outlined in Schemes 4.2 and 4.3. The bromide $\mathbf{6}.4PF_6$ (Scheme 4.2) was obtained in 61% yield by reacting the alcohol $\mathbf{5}.4PF_6$ ($\mathbf{B2}.4PF_6$) with carbon tetrabromide in dry MeCN in the presence of triphenylphosphine. Alkylation of the salt $\mathbf{7}.3PF_6$ with the bromide $\mathbf{6}.4PF_6$ in MeCN solution gave the dendrimer $\mathbf{B9}.18PF_6$ in a yield of 25% after counterion exchange. In Scheme 4.3, reaction of the bromide $\mathbf{6}.4PF_6$ with the salt $\mathbf{4}.2PF_6$, followed by counterion exchange, afforded the alcohol $\mathbf{9}.12PF_6$ in 65% yield.

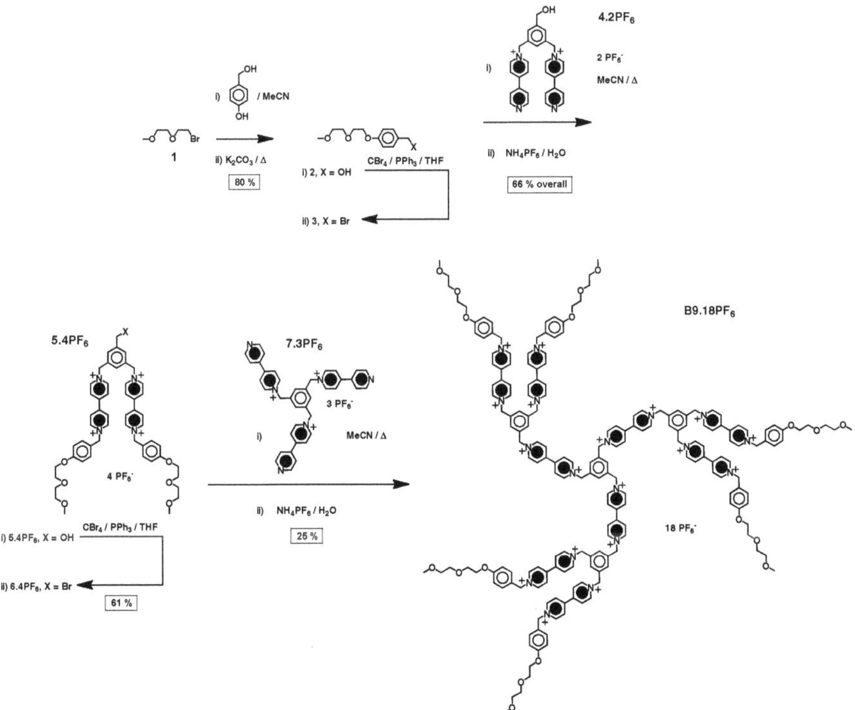

Scheme 4.2 Synthesis of dendron $\mathbf{B2}.4PF_6$ and dendrimer $\mathbf{B9}.18PF_6$

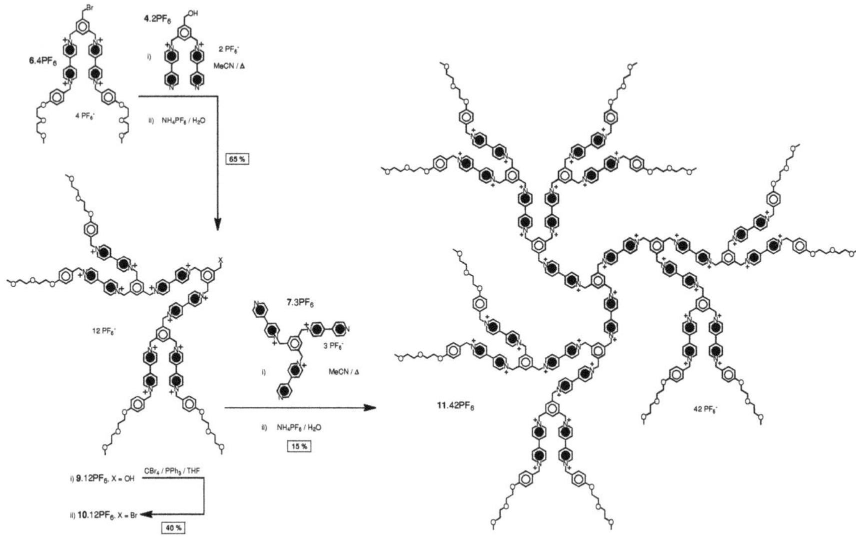

Scheme 4.3 Synthesis of dendrimer **B21**.42PF$_6$

The alcohol **9**.12PF$_6$ was treated with carbon tetrabromide in dry MeCN in the presence of triphenylphosphine to obtain the bromide, which was then reacted with the salt **7**.3PF$_6$ to give the dendrimer **B21**.42PF$_6$ in an overall yield of 15% after counterion exchange.

4.3 Absorption Spectra

Figure 4.1 shows the absorption spectra in MeCN of dendrimers **B9**$^{18+}$ and **B21**$^{42+}$ compared with the spectra of **dbV**$^{2+}$, taken as model compound of the dendrimer viologen units, at concentrations respectively 9 and 21 times higher than those of **B9**$^{18+}$ and **B21**$^{42+}$. It is evident that the absorption spectra of the dendrimers are not identical to that of the model compound, and that the differences are much more pronounced in the case of **B21**$^{42+}$. In particular, its spectrum is less intense than that of the model compound in the spectral region around the maximum and shows broad and weak absorption features, also present to a minor extent in the spectrum of **B9**$^{18+}$, that emerge from the low-energy tail of the intense UV band.

4.4 Eosin Complexation

It is well known that 4,4′-bipyridinium dications strongly interact with the dian-ionic form of eosin, **Ey**$^{2-}$, to yield charge-transfer (CT) complexes. [49, 50], We have previously observed that addition of **doV**$^{2+}$, as hexafluorophosphate salt, to a

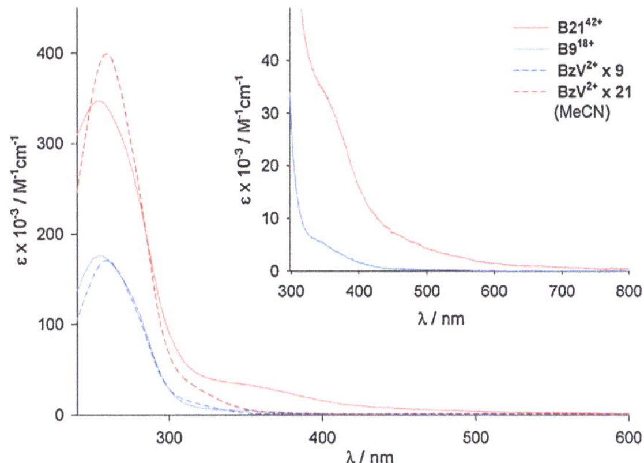

Fig. 4.1 Absorption spectra in MeCN of dendrimers **B9**$^{18+}$ (*blue line*) and **B21**$^{42+}$ (*red line*) compared with the spectra of 9 (*dashed blue line*) and 21 (*dashed red line*) **doV**$^{2+}$ units. *Inset*: Enlarged view ($\lambda > 300$ nm) of the spectra of the dendrimers

CH$_2$Cl$_2$ solution of **Ey**$^{2-}$, as tetrabutylammonium salt, causes (i) noticeable perturbations in the visible absorption band of **Ey**$^{2-}$, and (ii) complete quenching of its fluorescence because of the formation of a strong 1:1 complex ($K_{ass} > 10^6$ M^{-1}) [49].

Fluorescent titration experiments were performed to study the interaction between **Ey**$^{2-}$ and dendrimers **B9**$^{18+}$ and **B21**$^{42+}$. The experimental data obtained by titrating CH$_2$Cl$_2$ solutions of **B9**$^{18+}$ and **B21**$^{42+}$ (each containing ca. 1×10^{-5} M viologen units) with a solution of **Ey**$^{2-}$ (Fig. 4.2) show that the eosin fluorescence signal appears only when the number of added **Ey**$^{2-}$ exceeds the number of viologen units contained in each dendrimer. These results demonstrate that compounds **B9**$^{18+}$ and **B21**$^{42+}$ are capable of quenching the fluorescence of 9 and 21 eosin dianions, respectively, which means that each viologen unit of the dendrimers associates with one eosin dianion. Titration of dendron **B2**$^{4+}$ (9.0×10^{-6} M viologen units) confirms the stoichiometric formation of 1:1 complexes between the viologen units and eosin (Fig. 4.2). The quenching of the eosin fluorescence by association with the quencher (static mechanism) is

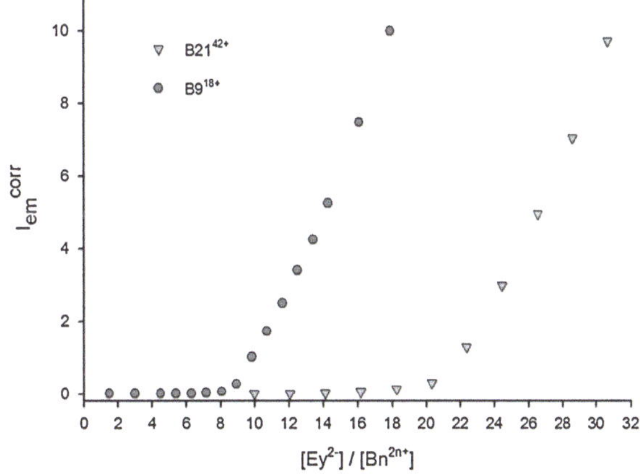

Fig. 4.2 Fluorescent titration experiments performed in CH$_2$Cl$_2$ solutions. Diagram of the intensity of the eosin **Ey**$^{2-}$ fluorescence band ($\lambda_{ex} = 500$ nm; $\lambda_{em} = 560$ nm) as a function of the [**Ey**$^{2-}$]/[**B**n^{2n+}] ratio: **B9**$^{18+}$ (*circles*); **B21**$^{42+}$ (*triangles*)

confirmed by the fact that dynamic quenching can be ruled out considering the short lifetime ($\tau = 3.8$ ns) of the eosin excited state and the low concentration ($< 5 \times 10^{-5}$ M) of the quencher.

4.5 Electrochemical Reduction

Electrochemical experiments were carried out in argon-purged MeCN solutions. The half-wave potential values, the number of exchanged electrons, and the diffusion coefficients of $\mathbf{dbV^{2+}}$ and dendrimers $\mathbf{B9^{18+}}$ and $\mathbf{B21^{42+}}$ are listed in Table 4.1, where the results previously obtained [18] for dendrimers $\mathbf{A9^{18+}}$ and $\mathbf{A21^{42+}}$ are also reported for comparison purposes (see Sect. 4.7).

Compounds $\mathbf{dbV^{2+}}$, $\mathbf{B9^{18+}}$, and $\mathbf{B21^{42+}}$ exhibit the two reduction processes typical of viologens. Cyclic voltammetric patterns show that the reducible viologen units of each dendrimer are equivalent and that the first reduction is reversible with nernstian behavior in all cases (Fig. 4.3), whereas the second process (particularly for $\mathbf{B21^{42+}}$) is affected by adsorption of the reduced species on the electrode.

The diffusion coefficients and the number of exchanged electrons have been measured (Table 4.1) from chronoamperometric experiments performed at the potential value of the first reduction process by using ultramicroelectrodes [51]. As expected, the diffusion coefficient of $\mathbf{B9^{18+}}$ is larger than that of $\mathbf{B21^{42+}}$. For both dendrimers the number of exchanged electrons is smaller than that expected on the basis of the number of the viologens contained in the structure. Apparently, only the external viologen units, 6 for $\mathbf{B9^{18+}}$ and 12 for $\mathbf{B21^{42+}}$, can be reduced in electrochemical experiments.

Unfortunately spectroelectrochemical experiments aimed at obtaining the absorption spectra of the reduced compounds could not be performed because of adsorption of the reduced species on the platinum minigrid used as working electrode.

Table 4.1 Half-wave reduction potentials, numbers of exchanged electrons, and diffusion coefficients in argon-purged MeCN solution, 298 K

	$E^{1a}_{1/2}$ (V *vs* SCE)	$E^{2a}_{1/2}$ (V *vs* SCE)	n^{b}_{el}	n^{c}	$10^{5} \times D$ (cm^2 s^{-1})
dbV^{2+}	−0.35	−0.77	1	1	160
B9^{18+}	−0.30	−0.78^{d}	6	9	45
B21^{42+}	−0.30	−0.73^{d}	11	21	27
A9^{18+e}	−0.29	−0.75^{d}	5	9	32
A21^{42+e}	−0.30	−0.76^{d}	14	21	27

[a] Working electrode glassy carbon, supporting electrolyte TBAPF$_6$
[b] Number of exchanged electrons obtained by chronoamperometric experiments (estimated error ±20%)
[c] Overall number of viologen units present in the compound as confirmed by the eosin complexation experiments
[d] Process affected by adsorption
[e] From Ref. [18]

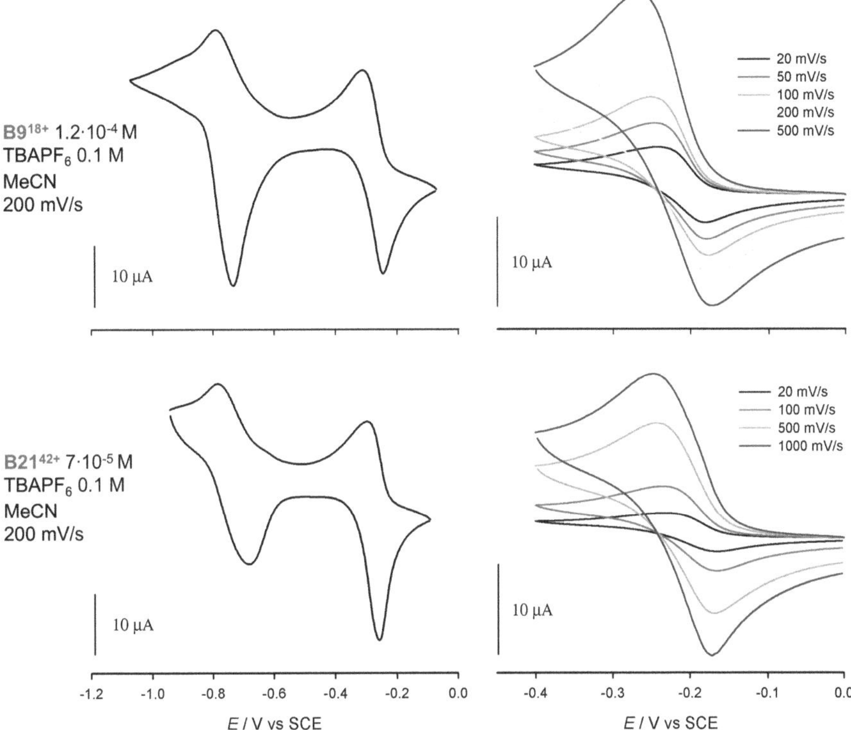

Fig. 4.3 Cyclic voltammetric behavior as a function of sweep rate for the first and second reduction process of dendrimer **B9**$^{18+}$ (1.2 × 10^{-4} M) and dendrimer **B21**$^{42+}$ (7 × 10^{-5} M). Argon-purged MeCN solutions, working electrode glassy carbon, supporting electrolyte TBAPF$_6$ (0.1 M)

4.6 Photosensitized Reduction

It is well known that one-electron reduction of viologen compounds can be conveniently performed by using suitable photosensitizers in the presence of sacrificial reductants. Light excitation leads the photosensitizer to a long-lived excited state that can transfer an electron to a viologen during an encounter. The back electron-transfer reaction between the oxidized photosensitizer and the reduced viologen is prevented by fast reduction of the oxidized photosensitizer by the sacrificial reductant. This kind of photoinduced processes have been extensively exploited for photogeneration of hydrogen from aqueous solutions ([15], see, e.g. [52, 53]), as well as to power artificial molecular devices and machines [54–63].

Previous investigations [18] showed that the photochemical reduction of viologen compounds can be efficiently performed in degassed CH$_2$Cl$_2$ solution by using 9-methylanthracene [64] as a photosensitizer and triethanolamine (TEOA) as a sacrificial reductant. As schematically shown in Fig. 4.4, light excitation of

Fig. 4.4 Scheme of doV^{2+} photosensitized reduction

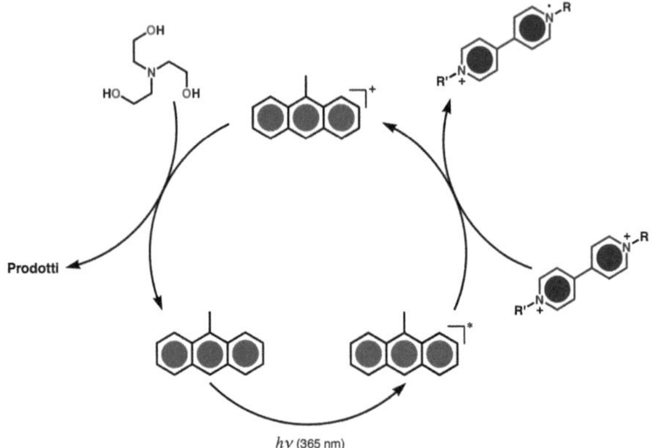

Table 4.2 Ratio between dimerized (V_d^+) and monomeric (V_m^+) monoreduced viologen units, percentages of the total monoreduced viologen units (V_{tot}^+) and numbers of exchanged electrons (n_{phot}) at the photostationary state

	V_d^+/V_m^+	% V_{tot}^+	n_{phot}[a]	n[b]
doV^{2+}	0	100	10	1
$B2^{4+}$	15	75	15	2
$B9^{18+}$	132	53	47	9
$B21^{42+}$	110	41	85	21
$A2^{4+}$	21	72	14	2
$A9^{18+c}$	85	48	43	9
$A21^{42+c}$	82	60	126	21

Degassed CH_2Cl_2 solution, irradiation with 365-nm light, 9-methylanthracene 1.2×10^{-4} M, TEOA 5.0×10^{-2} M

[a] Estimated error $\pm20\%$

[b] Overall number of viologen units present in compounds as confirmed by the eosin complexation experiments

[c] Adjusted from Ref. [18]

9-methylanthracene in CH_2Cl_2 solution in the presence of TEOA causes the reduction of doV^{2+} to its monomeric radical cation $doV^{\cdot+}$ (Table 4.2, inset of Fig. 4.5). Figure 4.5 shows the spectral changes observed upon irradiation with 365-nm light of a degassed CH_2Cl_2 solution containing 9-methylanthracene (1.2×10^{-4} M), TEOA (5.0×10^{-2} M), and $B21^{42+}$ (4.0×10^{-6} M).

The spectral features that arise upon irradiation are characteristics of the formation of both monomeric and dimeric reduced viologen species; similar changes were obtained for dendrimer $B9^{18+}$. The total number of electrons exchanged by dendrimers $B9^{18+}$ and $B21^{42+}$ at the photostationary state (Table 4.2), determined from absorbance measurements, is slightly lower than that obtained in chrono-amperometric experiments. This discrepancy can be attributed to small amounts of oxidizing impurities contained in the solvent and/or the samples used or, most

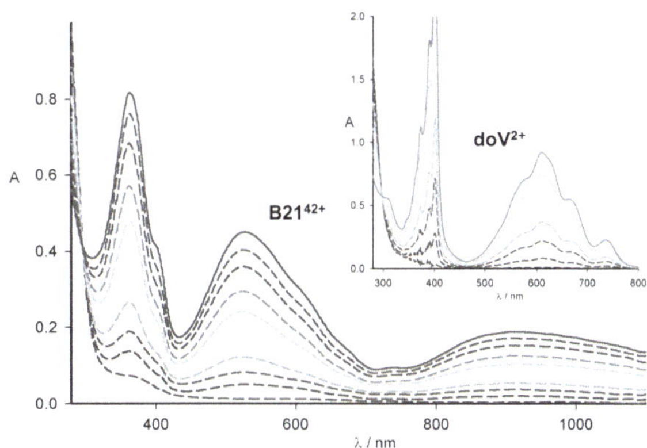

Fig. 4.5 Spectral changes (optical path $= 1$ cm) observed upon irradiation with 365-nm light of a degassed CH_2Cl_2 solution containing 1.2×10^{-4} M of 9-methylanthracene, 5.0×10^{-2} M of TEOA and 4.0×10^{-6} M of **B21**$^{42+}$ or (*inset*) 6.7×10^{-5} M of **doV**$^{2+}$. The *solid line* corresponds to the photostationary state. A solution containing 1.2×10^{-4} M of 9-methylanthracene was used as reference

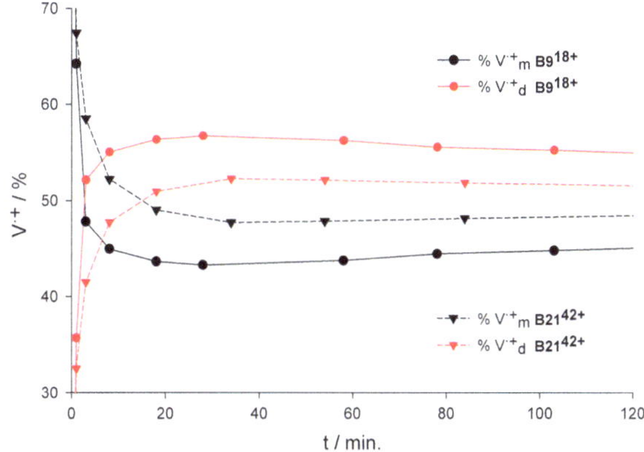

Fig. 4.6 Changes in the percentages of the monomeric (V_m^+, *black symbols*) and dimerized (V_d^+, *red symbols*) monoreduced viologen units on increasing irradiation time for the dendrimers **B9**$^{18+}$ (*circles*) and **B21**$^{42+}$ (*triangles*)

probably, to underestimation of the dimer concentration because of the ε values [65] used refer to a different solvent.

Spectra recorded at different irradiation time showed that the fraction of monomeric and dimeric reduced-viologen species changes with time (Fig. 4.6). At the photostationary state, dendrimer **B9**$^{18+}$ shows a prevalence of dimerized (ca. 57%) over monomeric (ca. 43%) viologen units; for dendrimer **B21**$^{42+}$, the fraction of dimerized viologen units is slightly lower.

For comparison purposes we have performed the photochemical reduction of dendrons **B2**$^{4+}$ and **A2**$^{4+}$ (Table 4.2) finding that (i) the number of reducible viologen units is slightly lower than expected and (ii) the percentage of dimerization is low and almost unaffected by the nature of the stoppers.

4.7 Discussion

The results obtained in this paper will now be discussed and compared with those previously reported [18, 49] for similar dendrimers terminated with bulky tetra-arylmethane groups (compounds $A9^{18+}$ and $A21^{42+}$, Tables 4.1 and 4.2).

4.7.1 Absorption Spectra

The absorption spectra of dendrimers $B9^{18+}$ and $B21^{42+}$ are different, particularly in the case of $B21^{42+}$, from the spectrum of dbV^{2+}. The differences can derive from the fact that (i) dbV^{2+} is not a fully satisfactory model, at least from a spectroscopic viewpoint, for the viologen units contained in the dendritic structure—each unit in the dendrimer shares benzyl groups with other units (Scheme 4.1) and some benzyl groups carry bismethyleneoxy substituents, and (ii) charge-transfer (CT) interactions originate between the electron-acceptor viologen units and the proximate (through bond) or remote (through space) electron-donor aryloxy units, as suggested by the appearance of an absorption tail for $\lambda > 320$ nm.

The previously studied dendrimers $A9^{18+}$ and $A21^{42+}$ exhibited a similar behavior; a quantitative analysis shows, however, that $A21^{42+}$ has an absorption derived by CT interactions that is less intense (about one half) and shifted towards lower energy compared to $B21^{42+}$. These results show that the terminal stoppers play some role in determining the spectroscopic properties of such dendrimers.

4.7.2 Host Properties Toward Eosin

We have found that both the **A**-type and **B**-type dendrimers are able to host a number of eosin dianions equal to the number of viologen units present in their branches. This result shows that neither the dendrimer generation nor the nature of the peripheral groups plays a role in the formation of the complexes between viologen and eosin moieties. Clearly, eosin anions can penetrate in the interior of the positively charged dendrimers, replacing the PF_6^- counter ions. The dendrimers can be viewed as polyvalent scaffolds with a well-defined number of independent seats where single eosin dianions can be hosted.

4.7.3 Charge Pooling

The electrochemical and photosensitization experiments performed on the **A**-type and **B**-type dendrimers have evidenced that (Tables 4.1 and 4.2) in all cases only a fraction of the viologen units can be reduced. Within the experimental error, this fraction corresponds to the number of the viologen units present in the outer shell (6 for $\mathbf{A9}^{18+}$ and $\mathbf{B9}^{18+}$, 12 for $\mathbf{A21}^{42+}$ and $\mathbf{B21}^{42+}$).

The fact that the number of reducible viologen units is smaller than expected cannot be attributed to lack of branches in the structures of the dendrimers because the results obtained for eosin complexation clearly show that the examined compounds do contain 9 ($\mathbf{A9}^{18+}$ and $\mathbf{B9}^{18+}$) and 21 ($\mathbf{A21}^{42+}$ and $\mathbf{B21}^{42+}$) viologen units, in agreement with the NMR and mass spectra characterization.

The *electrochemical reduction* experiments have shown that in each dendrimer, the first reduction process of *all* its reducible viologen units occurs at the same potential and that the first reduction process of *all* the reducible viologen units of *all* dendrimers occurs at the same potential. An interesting result is that the reduction potential of a reducible unit is not affected by the state of the other reducible units, which is an ideal property for a charge pooling system. Another result that emerges clearly from the data obtained is that the nature of the peripheral groups (tetraarymethane and aryloxy units for the **A**-type and **B**-type families, respectively) affects neither the number of reducible viologen units nor the first reduction potential. Two concomitant effects can be taken into account to explain the lack of complete electrochemical reduction: (i) upon reduction of the external viologen shell, the structure of the dendrimer shrinks, favoured by dimerization of the reduced units, thereby preventing the internal viologen units to "see" the electrode, and (ii) the internal viologens being engaged in tight ionic couples with the hexafluorophosphate counter-anions become more difficult to reduce, thereby preventing electron-hopping from external to internal viologen units.

The *photosensitized reduction* experiments have shown that the numbers of viologen units reducible under the photochemical conditions are in fair agreement with those obtained by chronoamperometric experiments, confirming that only the viologens of the external shell can be reduced. The photosensitized experiments have also revealed that formation of the one-electron reduced viologen units is accompanied by their dimerization reaction. The lack of reduction of all the viologen units in such experiments can be explained considering that (i) the uncharged photosensitizer cannot displace the counter-ions assembled in proximity of the highly charged core of the dendrimer, and (ii) the interaction between the photosensitizer and the internal viologen units can be prevented by the dimerization of the external reduced units that shrinks the dendrimer structure. We have also evidenced that dimerization is not a strongly favoured process as shown by the fact that it does not occur for the monoviologen $\mathbf{doV}^{2+}$ species and leads to only 15–20% of associated species in the case of dendrons $\mathbf{B2}^{4+}$ and $\mathbf{A2}^{4+}$ which are structurally preorganized to give dimerization (Table 4.2). In the case of the dendrimers, the fraction of dimerized viologen units is higher (about 45 and 57% for the **A**-type and **B**-type families, respectively), as expected because of the possibility to give dimers also between reduced units belonging to different dendrons. The results also show that the bulky tetraarylmethane peripheral moieties disfavour formation of dimers compared with the smaller aryloxy groups.

4.8 Conclusion

Continuing our work on polyviologen dendrimers we have found that these compounds independently of the nature and bulkiness of the terminal groups ($A9^{18+}$, $A21^{42+}$, $B9^{18+}$, and $B21^{42+}$) can act as polytopic receptors toward electron-donor substrates hosting a number of eosin dianions equal to the number of viologen units present in their branches. The data obtained clearly show that the host–guest interactions, that drive complex formation in low polar medium, are not affected by either the steric hindrance of the terminal groups or the electronic interactions that such groups establish with the viologen units. These results are of interest for the design and construction of dendrimers capable of performing functions related to drug delivery or sensing applications.

In principle, polyviologen dendrimers can behave as molecular batteries [16–31] being potentially capable of storing at easy accessible potential values a number of electrons twice that of the viologen units. Although it has been reported [16, 17] that in dendrimers very similar to those here described all the viologen units are reducible, our results clearly show that only a fraction of such units can be reduced and throw light on the reasons why charge-pooling is incomplete and on the role played by the terminal groups.

The fact that the number of reducible viologen units is smaller than expected cannot be attributed to lack of branches in the structures of the dendrimers because the results obtained for eosin complexation clearly show that the examined compounds do contain 9 ($A9^{18+}$ and $B9^{18+}$) and 21 ($A21^{42+}$ and $B21^{42+}$) viologen units.

Photosensitized reduction experiments have evidenced that for both the **A**-type and **B**-type families of dendrimers a fraction of the monoreduced viologen units undergoes dimerization and that such a process prevails for the dendrimers in which the bulky tetraarylmethane terminal groups have been removed.

We have also evidenced a clear dendrimer effect on the dimerization process involving the reduced viologen units. The dendritic structure of compounds $B9^{18+}$, $B21^{42+}$, $A9^{18+}$, and $A21^{42+}$ provides an environment capable of favouring dimerization because it forces the reduced viologen units to occupy close positions and is flexible enough to enable the interactions not only between the reduced units belonging to the same branch (intra-dendron interactions), but also between the reduced units contained in different branches (inter-dendron interactions).

References

1. Frechet JM, Tomalia DA (2001) Dendrimers and other dendritic polymers. Wiley, New York
2. Newkome GR, Vögtle F (2002) Dendrimers and dendron. Wiley-VCH, Weinheim
3. Vögtle F, Richardt G, Werner N (2007) Dendritische Moleküle. Teubner, Stuttgart
4. Chase PA, Gebbink RJMK, van Koten G (2004) J Organomet Chem 689:4016–4054
5. Ong W, Gomez-Kaifer M, Kaifer AE (2004) Chem Commun 1677–1683
6. Ballauff M, Likos CN (2004) Angew Chem Int Ed 43:2998–3020

7. Caminade AM, Majoral JP (2004) Acc Chem Res 37:341–348
8. Tomalia DA (2005) Prog Polym Sci 30(294–324), special issues on dendrimers and dendritic polymers (eds: Tomalia DA, Frechet JM)
9. Scott RWJ, Wilson OM, Crooks RM (2005) J Phys Chem B 109:692–704
10. Mery D, Astruc D (2006) Coord Chem Rev 250:1965–1979
11. Mery D, Astruc D (2007) New J Chem 31(7)101-123, special issue on dendrimers (ed: Majoral JP)
12. Ceroni P, Bergamini G, Marchioni F, Balzani V (2005) Prog Polym Sci 30:453–473
13. De Schryver FC, Vosch T, Cotlet M, Van der Auweraer M, Müllen K, Hofkens J (2005) Acc Chem Res 38:514–522
14. Balzani V, Credi A, Venturi M (2008) Molecular devices and machines: concepts and perspectives for the nanoworld. Wiley-VCH, Weinheim Ch. 6
15. Balzani V, Credi A, Venturi M (2008) ChemSusChem 1. doi:10.1002/cssc.200700087
16. Heinen S, Walder L (2000) Angew Chem Int Ed 39:806–809
17. Heinen S, Meyer W, Walder L (2001) J Electroanal Chem 498:34–43
18. Marchioni F, Venturi M, Ceroni P, Balzani V, Belohradsky M, Elizarov AM, Tseng HR, Stoddart LF (2004) Chem Eur J 10:6361–6368
19. Ardoin N, Astruc D (1995) Bull Soc Chim Fr 132:875–909
20. Bryce MR, Devonport W (1996) In: Newkome GR (ed) Advances in dendritic macromolecules, vol 3. JAI Press, London, pp 115–149
21. Gorman C (1998) Adv Mater 10:295–309
22. Balzani V, Campagna S, Denti G, Juris A, Serroni S, Venturi M (1998) Acc Chem Res 31: 26–34
23. Cuadrado I, Morán M, Casado CM, Alonso B, Losada J (1999) Coord Chem Rev 193–195:395
24. Nielsen MB, Lomholt C, Becher J (2000) Chem Soc Rev 29:153–164
25. Astruc D (2000) Acc Chem Res 33:287–298
26. Serroni S, Campagna S, Puntotiero F, Di Pietro C, McClenaghan ND, Loiseau F (2001) Chem Soc Rev 30:367–375
27. Alonso B, Astruc D, Blais J-C, Nlate S, Rigaut S, Ruiz J, Sartor V, Valério C (2001) CR Acad Sci Paris Chimie 4:173–180
28. Juris A, Venturi M, Ceroni P, Balzani V, Campagna S, Serroni S (2001) Collect Czech Chem Commun 66:1–32
29. Juris A (2001) In: Balzani V (ed) Electron transfer in chemistry, vol 3. Wiley-VCH, Weinheim, pp 655–714
30. Astruc D, Chardac F (2001) Chem Rev 101:2991–3031
31. Gorman CB (2003) CR Acad Sci Paris Chimie 6:911–918
32. Valério C, Fillaut JL, Ruiz J, Guittard J, Blais JC, Astruc D (1997) J Am Chem Soc 119:2588–2589
33. Balzani V, Ceroni P, Gestermann S, Kauffmann C, Gorka M, Vögtle F (2000) Chem Commun 853–854
34. Gong LZ, Hu QS, Pu L (2001) J Org Chem 66:2358–2367
35. Xu MH, Lin J, Hu QS, Pu L (2002) J Am Chem Soc 124:14239–14246
36. Astruc D, Daniel M-C, Ruiz J (2004) Chem Commun 2637–2649
37. Baars MWPL, Meijer EW (2000) Top Cur Chem 210:131–182
38. Astruc D, Chardac F (2001) Chem Rev 101:2991–3024
39. Schalley CA, Baytekin B, Baytekin HT, Engeser M, Felder T, Rang A (2006) J Phys Org Chem 19:479–490
40. Darbre T, Reymond J-L (2006) Acc Chem Res 39:925–934
41. van Heerbeek R, Kamer PCJ, van Leeuwen PNMW, Reek JNH (2006) Org Biomol Chem 4:211–223
42. Hahn U, Kaufmann A, Nieger M, Julinek O, Urbanova M, Vögtle F (2006) Eur J Org Chem 1237–1244
43. Gingras M, Raimundo J-M, Chabre YM (2007) Angew Chem Int Ed 46:1010–1017

44. Chai M, Holley AK, Kruskamp M (2007) Chem Commun 168–170
45. Puntoriero F, Ceroni P, Balzani V, Bergamini G, Vögtle F (2007) J Am Chem Soc 129:10714–10719
46. Summers LA (1980) The bipyridinium herbicides. Academic Press, London
47. Bird CL, Kuhn AT (1981) Chem Soc Rev 10:49–82
48. Monk PMS (1998) The viologens: physicochemical properties, synthesis, and applications of the salts of 4, 4′-bipyridine. Wiley, Chichester
49. Marchioni F, Venturi M, Credi A, Balzani V, Belohradsky M, Elizarov AM, Tseng HR, Stoddart JF (2004) J Am Chem Soc 126:568–573
50. Willner I, Eichen Y, Rabinovitz M, Hoffman R, Cohen S (1992) J Am Chem Soc 114:637–644
51. Denuault G, Mirkin MV, Bard AJ (1991) J Electroanal Chem 308:27–38
52. Amouyal E (1995) Sol Energy Mater Sol Cells 38:249–276
53. Balzani V, Credi A, Venturi M (2008) Molecular devices and machines: concepts and perspectives for the nanoworld. Wiley-VCH, Weinheim Ch. 7 and references therein
54. Balzani V, Credi A, Raymo FM, Stoddart JF (2000) Angew Chem Int Ed 39:3349–3391
55. Ballardini R, Balzani V, Credi A, Gandolfi MT, Venturi M (2001) Acc Chem Res 34:445–455
56. Balzani V (2003) Photochem Photobiol Sci 2:459–476
57. Balzani V, Credi A, Ferrer B, Silvi S, Venturi M (2005) Top Curr Chem 262:1–27
58. Balzani V, Credi A, Silvi S, Venturi M (2006) Chem Soc Rev 35:1135–1149
59. Browne WR, Feringa BL (2006) Nat Nanotechnol 1:25–35
60. Ballardini R, Credi A, Gandolfi MT, Marchioni F, Silvi S, Venturi M (2007) Photochem Photobiol Sci 6:345–356
61. Kay ER, Leigh DA, Zerbetto F (2007) Angew Chem Int Ed 46:72–191
62. Pollard MM, Klok M, Pijper D, Feringa BL (2007) Adv Funct Mater 17:718–729
63. Balzani V, Credi A, Venturi M (2008) Molecular devices and machines: concepts and perspectives for the nanoworld. Wiley-VCH, Weinheim
64. Montalti M, Credi A, Prodi L, Gandolfi MT (eds) (2006) Handbook of photochemistry, 3rd edn. CRC Taylor & Francis, Boca Raton
65. Geuder W, Hunig S, Suchy A (1986) Tetrahedron 42:1665–1677

Chapter 5
Dendrimers: Tweezering the Core of Dendrimers: Medium Effect on the Kinetic and Thermodynamic Properties

5.1 Introduction

Molecular recognition is defined by the information and the energy involved in selection and binding of substrate(s) by a given receptor molecule [1]. Molecular recognition phenomena imply the formation of supramolecular species characterized by peculiar structural, thermodynamic, and kinetic features determined by molecular information stored in receptor and substrate(s) that lead to selective binding. Stability of the supramolecular host–guest complex substantially depends on the medium and results from a subtle balance between solvation [2] (of both receptor and substrate) and complexation (i.e., "solvation" of the substrate by the receptor) [1].

Dendrimers ([3]; dendrimers and other dendritic polymers [4]) are repeatedly branched molecules that can contain selected functional groups in predetermined sites of their multiarmed structure. In spite of their large structures, that usually make dendrimers attractive as host molecules, suitably designed dendrimers can be involved as guests in molecular recognition phenomena (for some recent examples, see [5–12]). In such cases, the host species does not interact with the whole

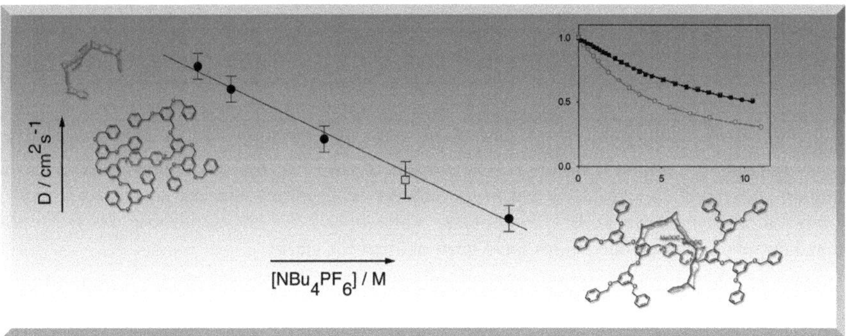

M. Baroncini, *Design, Synthesis and Characterization of new Supramolecular Architectures*, Springer Theses, DOI: 10.1007/978-3-642-19285-2_5,
© Springer-Verlag Berlin Heidelberg 2011

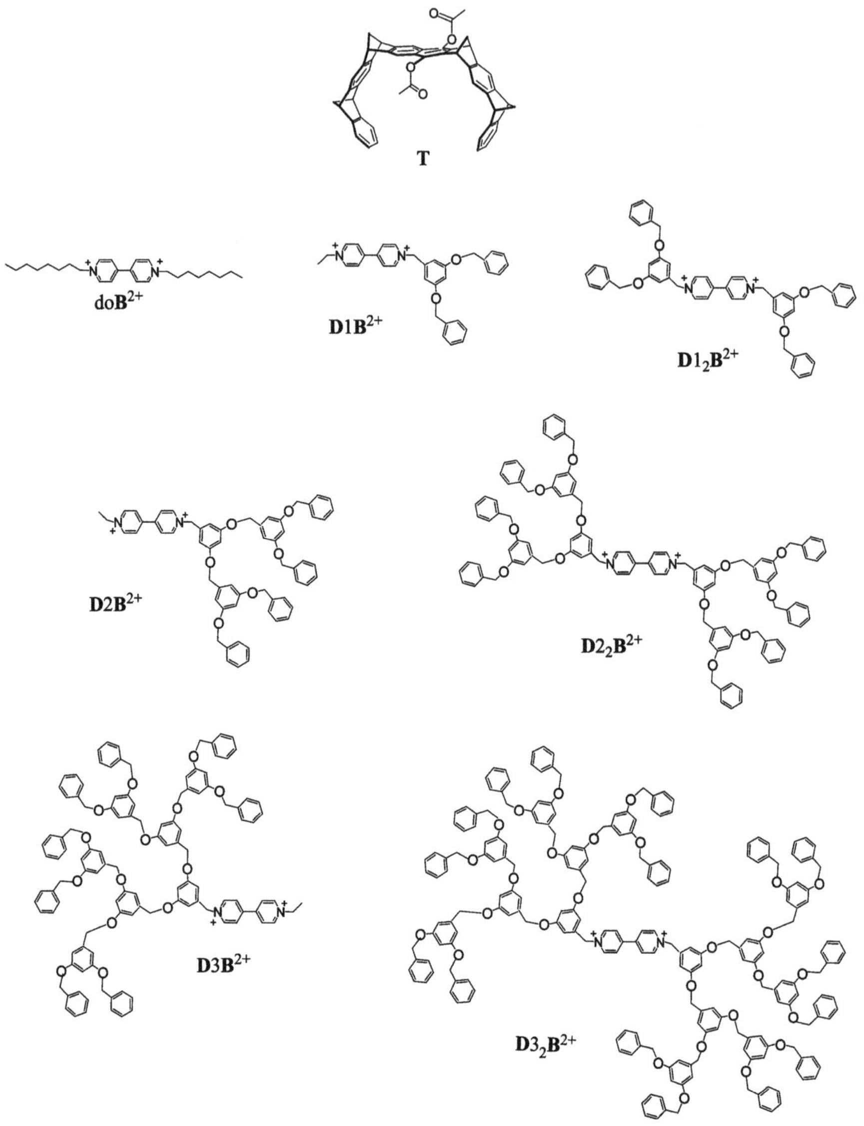

Scheme 5.1 Formulas of the investigated compounds and abbreviations used. The tweezer-shaped host is **T**; the dendrimers are generically indicated as $\mathbf{D}n_m\mathbf{B}^{2+}$, where $\mathbf{B}^{2+}$ stands for the 4,4′-bipyridinium core and **D** for the 1,3-dimethyleneoxybenzene-based dendrons; n indicates the dendron generation ($n = 1, 2, 3$) and m the number of dendrons attached to the core: $m = 1$ stands for nonsymmetric $\mathbf{D}n\mathbf{B}^{2+}$ dendrimers, while $m = 2$ for the symmetric $\mathbf{D}n_2\mathbf{B}^{2+}$ ones. Hexafluorophosphate counteranions have been omitted for clarity

dendritric structure, but only with specific component units. In particular, we have previously reported [13, 14] that dendrimers containing an electron-acceptor 4,4′-bipyridinium core [15, 16] appended with one ($\mathbf{D}n\mathbf{B}^{2+}$) or two ($\mathbf{D}n_2\mathbf{B}^{2+}$) polyaryl-ether dendrons (Scheme 5.1) are known to function as guests in molecular recognition phenomena involving a concave electron-donor molecule comprising a naphthalene and four benzene components bridged by four methylene groups [17–19] (tweezer T in Scheme 5.1). The Dn_mB^{2+} dendrimers show an amphiphilic character since they are constituted of a hydrophilic 4,4′-bipyridinium core surrounded by hydrophobic 1,3-dimethyleneoxybenzene-based dendrons. Therefore, the medium may affect the shape of the dendrimers.

Here we examine the effects induced by a change in solvent polarity and addition of tetrabutylammonium hexafluorophosphate, hereafter indicated as NBu_4PF_6, on the thermodynamic and kinetic features of the host–guest complex formation between the Dn_mB^{2+} substrates and the endoreceptor tweezer T. Characterization of host–guest complexes based on charge-transfer interactions often combines spectroscopic and electrochemical studies, but conventional electrochemical techniques require the presence of supporting electrolytes that may affect the stability and the kinetics of complex formation, especially if host and/or guest are charged species. In this regard, it has been shown that redox processes in nonaqueous solutions may be very sensitive to changes in solvents and supporting electrolytes [20]. Furthermore, the presence of salts has been demonstrated to affect electron-transfer processes in supramolecular adducts formed by donor/acceptor interactions (ion–radical pairs are formed upon photoinduced electron transfer of the corresponding ground state charge-transfer complex [21, 22]; both charge- and electron-transfer processes occur at the ground state [23]).

5.2 Photophysical and Electrochemical Properties of Tweezer T

In CH_2Cl_2 solution, T shows relatively weak, low-energy absorption bands, and a strong fluorescence band ($\lambda_{max} = 344$ nm, $\tau = 9.5$ ns, $\Phi = 0.53$) typical of the naphthalene chromophoric group. In CH_2Cl_2/CH_3CN 9:1 (v/v) solution, T shows an irreversible oxidative process at +1.6 V, whereas no reduction process has been observed in the potential window of the solvent used (up to -2.0 V $vs.$ SCE).

5.3 Photophysical and Electrochemical Properties of Dendrimers Dn_mB^{2+}

The investigated dendrimers contain three types of chromophoric groups, namely a 4,4′-bipyridinium unit at the core, 1,3-dimethyleneoxybenzene units in the branches, and benzene units at the periphery. Dendrimers exhibit a strong absorption band in the UV region that does not coincide with the summation of the spectra of the component units, particularly because of the presence of a broad and weak

Table 5.1 Half-wave potentials (V vs. SCE) in CH_2Cl_2/CH_3CN 9:1 (v/v), NBu_4PF_6 0.1 M, except as otherwise noted

	$B^{2+} \rightarrow B^{\cdot+}$	$B^{\cdot+} \rightarrow B$
$D1B^{2+}$	−0.29	−0.77
$D2B^{2+}$	−0.27	−0.73
$D3B^{2+}$	−0.27	−0.73
$D1_2B^{2+a}$	−0.25	−0.73
$D2_2B^{2+}$	−0.24	−0.72
$D3_2B^{2+}$	−0.24	−0.72

[a] CH_2Cl_2/CH_3CN 3:1 (v/v)

absorption tail at $\lambda > 300$ nm that is assigned to a charge-transfer transition from 1,3-dimethyleneoxybenzene electron-donor units to the 4,4′-bipyridinium electron-acceptor core. Dimethyleneoxybenzene and, accordingly, Frechet-type dendrons [24] are known to exhibit fluorescence ($\lambda_{max} = 350$ nm and $\tau < 1$ ns). Dendrimers, instead, are not fluorescent because the excited state localized on the 1,3-dimethyleneoxybenzene units is deactivated via a fast electron-transfer process to the 4,4′-bipyridinium, as revealed by transient absorption measurements (P Ceroni et al, unpublished results). The 4,4′-bipyridinium core, B^{2+}, is a well-known electroactive unit that undergoes two successive, reversible, one-electron reduction processes at easily accessible potentials that correspond to the formation of a radical cation ($B^{2+} \rightarrow B^{\cdot+}$) and a neutral ($B^{\cdot+} \rightarrow B$) species [13, 14]. In agreement with these expectations, the DnB^{2+} nonsymmetric dendrimers, as well as the Dn_2B^{2+} symmetric ones show two reversible one-electron transfer processes. The half-wave potential values ($E_{1/2}$) observed for reduction of the six dendrimers, gathered in Table 5.1, show that the 4,4′-bipyridinium electrochemical behavior is slightly affected by the presence of the dendritic branches: (i) the reduction potentials are slightly shifted toward negative potentials (30–40 mV) for the nonsymmetric dendrimers compared to the symmetric ones and (ii), within each family, the first generation dendrimer shows a slight negative shift of the potentials for both the reduction processes compared to the second and third generation dendrimers. The heterogeneous rate of electron transfer to the electrode surface is high for all the dendrimers. For example, in the case of $(D3)_2B^{2+}$ the two reduction processes show a Nernstian behavior at scan rates up to 5 V/s, thereby indicating no significant inhibition or site isolation effect on the dendrimer core by the dendrons.

5.4 ^{1}H NMR Spectra of Dendrimers in Different Media

As an example the case of the symmetric second generation dendrimer is discussed in detail. ^{1}H NMR spectra of 0.4 mM $D2_2B^{2+}$ recorded in CD_3CN (Fig. 5.1c) or CD_3COCD_3 [25] at 295 K are characterized by sharp and well-resolved resonances. On the other hand, the ^{1}H NMR spectrum of 0.4 mM $D2_2B^{2+}$ in CD_2Cl_2/CD_3CN 9:1 (v/v) at 295 K (Fig. 5.1a) shows broad resonances for inner protons of the 4,4′-bipyridinium core, while aromatic protons at the periphery of the dendrons

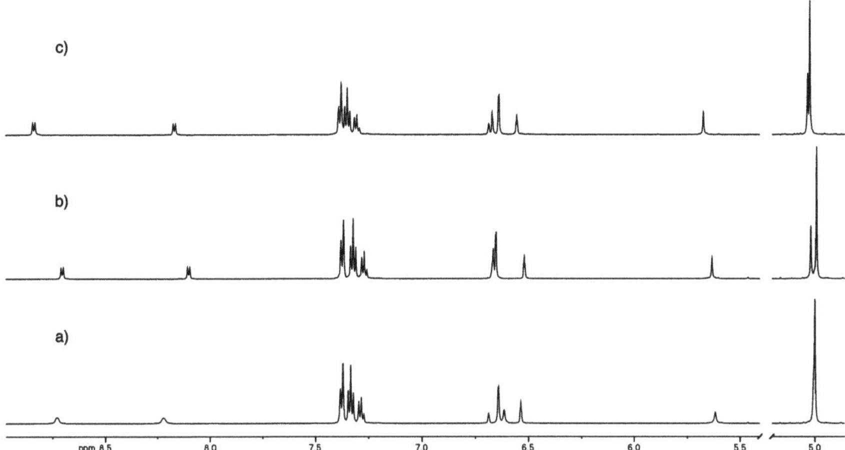

Fig. 5.1 ^{1}H NMR spectra (600 MHz) at 295 K of $D2_2B^{2+}$ 0.4 mM in **a** CD_2Cl_2/CD_3CN 9:1 (v/v), **b** CD_2Cl_2/CD_3CN 9:1 (v/v) and 0.15 M NBu_4PF_6, and **c** CD_3CN

are well resolved. (The line broadening of the signals belonging to the bipiridyl core is mainly due to a very short longitudinal relaxation time T_1. This finding may reflect the different solvation of the core in the two different solvents, i.e., CD_2Cl_2 and CD_3CN.) This behavior can be explained by many different processes, such as change in dendrimer conformation (dendron folding/unfolding), formation of aggregates [25], and ion pairing with PF_6^- counterions. The last hypothesis can be ruled out since, on the basis of the results obtained in the association with the tweezer (see below), the bipyridinium core and the PF_6^- counterions form a tight ion pair. To discriminate between intradendrimer (dendron folding/unfolding) and interdendrimer [e.g., formation of aggregates between bipyridinium cores (a somehow related molecule consisting of three arms each containing 4,4′-bi-pyridinium units attached covalently to a 1,3,5-trisubstituted benzene central core and each bearing at its other end a bulky hydrophobic tetraarylmethane-based terminating group revealed, in chloroform solution, the formation of aggregates, see: [26]) of different dendrimers] processes, NMR spectra of $D2_2B^{2+}$ at different concentrations have been recorded in CD_2Cl_2/CD_3CN 9:1 (v/v) solution at 295 K: no change has been observed at concentrations ranging between 1×10^{-4} and 5×10^{-3} M. It follows that formation of aggregates is unlikely to occur, especially in the dilute solutions ($\approx 10^{-5}$ M) used for fluorescence measurements. Indeed, normalized UV–vis absorption spectra of an air-equilibrated CH_2Cl_2/CH_3CN 9:1 (v/v) solution of $D2_2B^{2+}$ at 298 K coincide in the 1×10^{-6} to 5×10^{-4} M concentration range.

These findings rule out interdendrimer processes and point to a change in dendrimer conformation. In particular, a folding of dendrons around the dendritic core is likely to occur in low-polarity solvent, as confirmed by other measurements (see below).

To obtain more insight on this conformational change we performed ^{1}H NMR measurements in the presence of increasing quantity of NBu_4PF_6 aimed at reproducing the electrochemical conditions. Indeed, under such conditions, CD_2Cl_2/CD_3CN 9:1 (v/v) with 0.1 M NBu_4PF_6 as supporting electrolyte, no significant inhibition or site isolation effect on the dendrimer core by the dendrons has been evidenced [13, 14].

Titration of a CD_2Cl_2/CD_3CN 9:1 (v/v) solution containing 0.4 mM $\mathbf{D2_2B^{2+}}$ with NBu_4PF_6 results in a gradual sharpening (see, e.g., Fig. 5.1b) of the resonances of the ^{1}H NMR spectra with peak shapes resembling those obtained in CD_3CN. In addition, ^{1}H NMR spectra of 0.4 mM $\mathbf{D2_2B^{2+}}$ in CD_3CN solution have been found almost insensitive to NBu_4PF_6 addition. These results clearly indicate that the NBu_4PF_6 salt cannot be considered inert. In low-polarity solvent it affects the environment that becomes similar to that experienced by the dendrimers in high-polarity solvents. As a consequence the dendrimer conformation changes, most probably leading to unfolding of the dendritic structure.

5.5 Diffusion Coefficients of Dendrimers Measured by NMR Experiments

Pulse gradient stimulated echo NMR techniques [27] have been used to gain information on the conformation of the dendrimers through their diffusion in different media. In the present case, DOSY experiments [28] have been performed on a CD_2Cl_2/CD_3CN 9:1 (v/v) solution containing 0.4 mM $\mathbf{D2_2B^{2+}}$ in the presence of increasing amounts of NBu_4PF_6 up to 0.15 M.

Decays of the signal corresponding to the Ar–C$\underline{H}_2$O protons at 5 ppm as a function of the pulsed gradient strength have been plotted (Fig. 5.2) according to the Stejskal–Tanner Eq. 5.1 [29]:

$$\ln\frac{I}{I_0} = -d^2g^2G^2(D - d/3)D \tag{5.1}$$

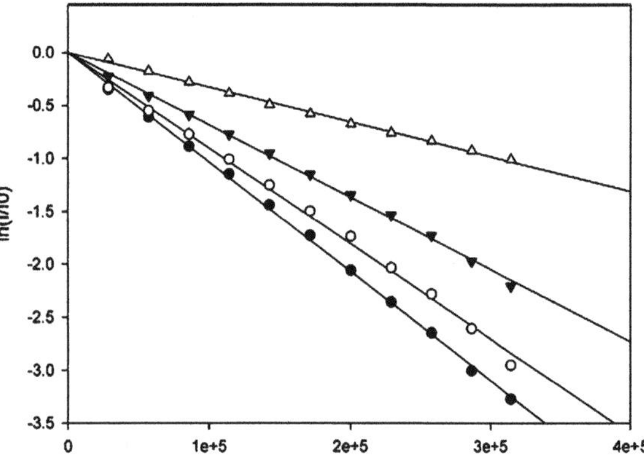

Fig. 5.2 Stejskal–Tanner plots of a CD_2Cl_2/CD_3CN 9:1 (v/v) solution containing 0.4 mM $\mathbf{D2_2B^{2+}}$ at 295 K in the presence of increasing amounts of NBu_4PF_6: 0 (*solid circles*), 15 (*open circles*), 60 (*solid triangles*), and 150 mM (*open triangles*)

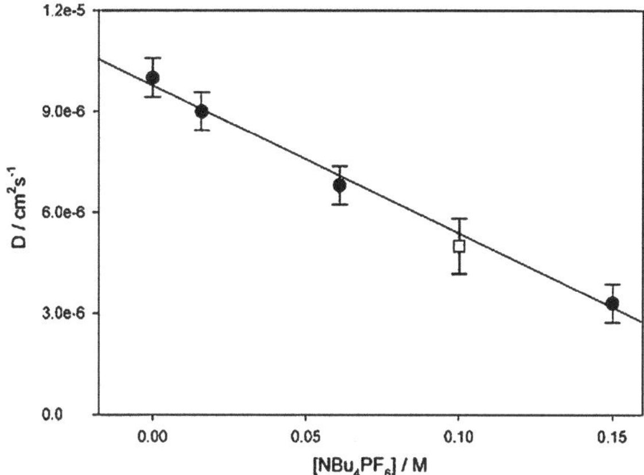

Fig. 5.3 Diffusion coefficient values of 0.4 mM **D2₂B²⁺** in air-equilibrated CD₂Cl₂/CD₃CN 9:1 (v/v) solution determined by DOSY experiments as a function of NBu₄PF₆ concentration (*solid circles*). *Open square* represents the diffusion coefficient value of **D2₂B²⁺** in argon-purged CH₂Cl₂/CH₃CN 9:1 (v/v) solution containing 0.1 M NBu₄PF₆, obtained by chronoamperometric experiments

where γ is the gyromagnetic ratio, δ is the duration of the pulse, G is the pulsed gradient strength, Δ is the time separation between the pulsed gradients, and D is the diffusion coefficient. The product $\delta^2\gamma^2G^2(\Delta - \delta/3)$ is termed the b value. The plot of $\ln(I/I_0)$ versus the b values for an isotropic solution gives a straight line, with a slope of $-D$.

In the case of the examined dendrimer (Fig. 5.2), Stejskal–Tanner plots are linear both in the absence and in the presence of NBu₄PF₆ and reveal that the diffusion coefficient of **D2₂B²⁺** decreases linearly with increasing salt concentration (Fig. 5.3): a more than threefold decrease is observed on going from 0 to 0.15 M NBu₄PF₆.

The diffusion coefficient D can be related to the hydrodynamic radius r_h by the Stokes–Einstein equation, derived by assuming that a spherical particle of colloidal dimension moves with uniform velocity in a fluid continuum. For compounds having a van der Waals volume much bigger than the volume of the solvent molecules, "sticking" boundary conditions are applicable, and the non-spherical form of the molecule has little influence since it moves together with solvent molecules [30]. Under these assumptions, the diffusion coefficient D is inversely proportional to the hydrodynamic radius, r_h, and the medium viscosity η, according to Eq. 5.2:

$$r_h = \frac{(kT)}{(6\pi\eta D)} \tag{5.2}$$

where k is the Boltzmann constant and T is the temperature.

By taking into account the medium viscosity that changes from 0.6759 mPa s for a CH₂Cl₂/CH₃CN 9:1 (v/v) mixture to 0.8060 mPa s for the same solvent mixture in the presence of 0.15 M NBu_4PF_6, the hydrodynamic radius of the dendrimer increases by a factor of 2.5. This value is quite large and can be partly due to the fact that the assumptions leading to Eq. 5.2 are not fully applicable to the present case.

Because the aggregate formation can be ruled out (vide supra), the observed trend has to be mainly attributed to an increase in dendrimer volume upon unfolding of the dendrons.

Steady-state cyclic voltammetric and chronoamperometric measurements with ultramicroelectrodes [31] provides an independent measure of diffusion coefficient for $\mathbf{D2_2B^{2+}}$ in CH_2Cl_2/CH_3CN 9:1 (v/v) in the presence of 0.1 M NBu_4PF_6 (open square in Fig. 5.3). Such a value is interpolated by the linear regression of the diffusion coefficients of $\mathbf{D2_2B^{2+}}$ in CH_2Cl_2/CH_3CN 9:1 (v/v) at different salt concentrations determined by using the DOSY technique (Fig. 5.3). The necessity to take into account the effect of supporting electrolyte addition in measuring diffusion coefficients has already been stressed by Kaifer et al. [32].

5.6 Electrochemical Measurements

Host–guest complex formation between tweezer $\mathbf{T}$ and $\mathbf{Dn_mB^{2+}}$ dendrimers is essentially driven by electron-donor/acceptor interactions between the electron-acceptor core of the dendrimers and the electron-donor cavity of the host. Clear evidence of the complex formation is given by the half-wave potential values ($E_{1/2}$), corresponding to the first one-electron reduction process of the 4,4′-bipyridinium dendritic cores that move toward more negative values upon host addition (compare open and filled circles of Fig. 5.4) because the electron-withdrawing character of the dendritic cores is reduced by the interaction with the electron-donor cavity of $\mathbf{T}$.

Both geometrical and interactional complementarity requirements are fulfilled in the formation of these host–guest complexes. Upon one-electron reduction of the 4,4′-bypiridinium dendritic core, however, the interactional complementarity between receptor $\mathbf{T}$ and monoreduced $\mathbf{Dn_mB^{+}}$ species is weakened. Thereby dissociation of the inclusion complexes occurs as demonstrated by the fact that the second reduction process of the bypiridinium core takes place at the same potential value in the absence and in the presence of $\mathbf{T}$ [13, 14].

5.7 Photophysical Measurements

Titrations of CH_2Cl_2 air-equilibrated solutions of $\mathbf{T}$ (ca. 15 μM) with solutions of $\mathbf{Dn_mB^{2+}}$ dendrimers and $\mathbf{doB^{2+}}$ model compound evidence a strong decrease of the $\mathbf{T}$ emission band. As an example, in Fig. 5.5 the data obtained for $\mathbf{D3_2B^{2+}}$ are reported (black lines). Time-resolved fluorescence measurements support a static

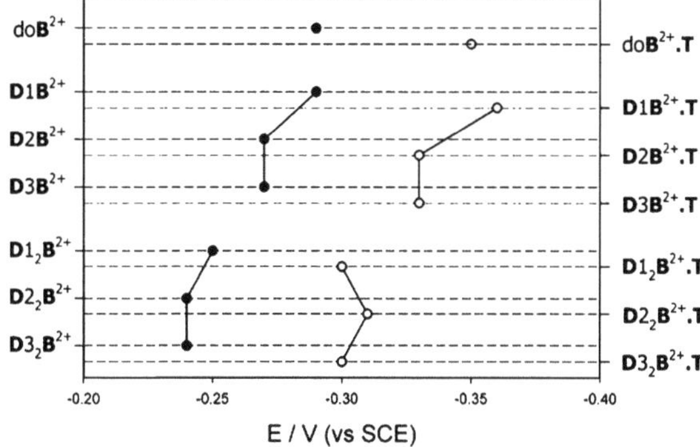

Fig. 5.4 Half-wave potentials for the first reduction processes of doB^{2+} model compound and Dn_mB^{2+} dendrimers (*filled circles*), and of their host–guest complexes with tweezer **T** (*open circles*) in argon-purged CH_2Cl_2/CH_3CN 9:1 (v/v) containing 0.1 M NBu_4PF_6. Under the experimental conditions used, more than 95% of the electroactive species is engaged in the complex formation. For solubility problems the $D1_2B^{2+}$ dendrimer and the $D1_2B^{2+}\cdot T$ complex have been studied in argon-purged CH_2Cl_2/CH_3CN 3:1 (v/v) solution containing 0.1 M NBu_4PF_6

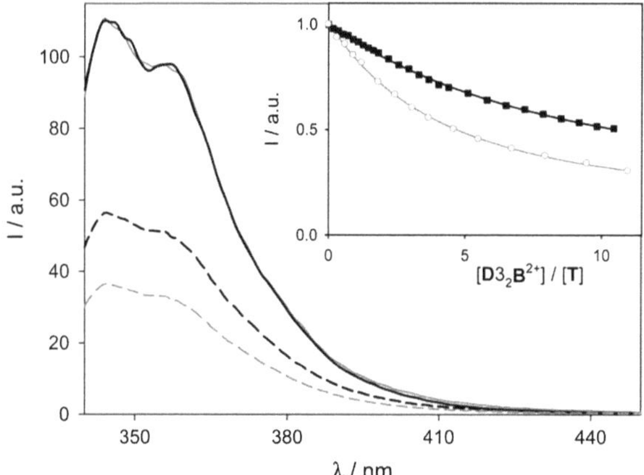

Fig. 5.5 Fluorescence spectra of a 16 µM air-equilibrated CH_2Cl_2 solution of **T** upon addition of 0 (*solid lines*) and 10 equiv (*dashed lines*) of $D3_2B^{2+}$ in the absence (*black lines*) and presence (*gray lines*) of 0.14 M NBu_4PF_6; $\lambda_{ex} = 334$ nm, 298 K. The inset shows the corrected emission intensities of **T** at 356 nm as a function of the added equivalents of $D3_2B^{2+}$, in the absence (*black squares*) and presence (*gray open circles*) of 0.14 M NBu_4PF_6. The *solid lines* show the corresponding fitting based on the formation of a 1:1 complex

quenching mechanism due to host–guest complex formation; the fluorescence lifetime of **T** is, indeed, not affected upon dendrimer addition, and there is no evidence of a double exponential decay. Therefore, the observed fluorescence intensity can be assigned to the uncomplexed host molecules [13, 14].

Table 5.2 Association constants for complexes between host T and Dn_mB^{2+} dendritic guests or doB^{2+} model compound in air-equilibrated CH_2Cl_2 and CH_2Cl_2/CH_3CN 1:3 (v/v) solution at 298 K

	K_{ass} (10^3 M^{-1})	
	CH_2Cl_2	CH_2Cl_2/CH_3CN 1:3 (v/v)
doB^{2+}	23	1.9
$D1B^{2+}$	34	2.4
$D2B^{2+}$	22	2.9
$D3B^{2+}$	16	2.2
$D1_2B^{2+}$	27	2.8
$D2_2B^{2+}$	18	2.9
$D3_2B^{2+}$	9.3	2.3

Fitting of the experimental fluorescence intensities is fully satisfactory ($r^2 > 0.99$ for the data shown in the inset of Fig. 5.5). The corresponding association constants (Table 5.2) are of the order of 10^4 M^{-1} in CH_2Cl_2 solution and a comparison of the obtained values evidences that (i) the complexes involving nonsymmetric DnB^{2+} dendrimers are more stable than those formed by the symmetric Dn_2B^{2+} analogues and (ii) upon increasing dendrimer generation, the stability of the complexes decreases within both (symmetric and nonsymmetric) dendrimer families. Therefore, in low-polarity solvent the dendritic branches disfavor, to some extent, the recognition of the 4,4′-bipyridinium core by the receptor. In CH_2Cl_2/CH_3CN 1:3 (v/v) solution the association constants are 1 order of magnitude lower and quite similar for all the investigated dendrimers (Table 5.2). This trend has also been observed by performing ^{1}H NMR measurements in pure CD_2Cl_2 and CD_2Cl_2/acetone-d$_6$ 1:2 (v/v) or $CDCl_3$/acetone-d$_6$ 1:2 (v/v).

The results of these investigations show that by increasing the polarity of the medium the stability of the complexes decreases and loses its dependence by the number and the generation of the dendrons appended to the 4,4′-bipyridinium core. Such experimental evidence is in agreement with the expectation because in a more polar solvent the electron donor–acceptor interactions are weaker, i.e., solvation of the hydrophilic dicationic core by higher polarity solvent molecules competes with "solvation" of the core by the host cavity more efficiently than in lower polarity solvent.

Ion pairing should be taken into account when adduct formation involves charged species and occurs in low dielectric media. In particular, if the formation of inclusion complexes involves previous ion-pair dissociation, the apparent stability constants are known to be concentration dependent [33]. In the present case, however, similar binding constant values have been obtained by independent titration experiments performed in air-equilibrated CH_2Cl_2 or CD_2Cl_2 solutions at 298 K and with concentration values that range between 10^{-5} to 10^{-4} M and 5×10^{-4} to 5×10^{-3} M by using the spectrofluorimetric technique and ^{1}H NMR measurements, respectively [13, 14]. It follows that complex formation does not require ion-pair dissociation and ion-paired adducts are formed:

$$Dn_mB^2 + (2PF_6^-) \leftarrow Dn_mB^{2+} + (2PF_6^-) \tag{5.3}$$

$$\mathbf{T} + \mathbf{D}n_m\mathbf{B}^2 + (2\mathrm{PF}_6^-) \rightleftarrows \mathbf{T} \cdot \mathbf{D}n_m\mathbf{B}^{2+}(2\mathrm{PF}_6^-) \tag{5.4}$$

Nevertheless, to simplify the representation of the dendrimers, the model compound, and their supramolecular adducts, PF_6^- counteranions have always been omitted.

To compare the results obtained by fluorescence and electrochemical experiments, the effect of $\mathrm{NBu_4PF_6}$ addition on the stability of the host–guest complexes in $\mathrm{CH_2Cl_2}$ solution have also been investigated. In particular, the behavior of $\mathbf{D1_2B^{2+}}$, $\mathbf{D3B^{2+}}$, and $\mathbf{D3_2B^{2+}}$ dendrimers and $\mathrm{do}\mathbf{B^{2+}}$ model compound have been examined in order to investigate the effect of the number and generation of the appended dendrons.

Upon addition of $\mathrm{NBu_4PF_6}$ (up to 0.14 M) to a $\mathrm{CH_2Cl_2}$ solution of $\mathbf{T}$, absorption and emission spectra of $\mathbf{T}$ do not show any appreciable change (compare spectra represented by black and gray solid lines in Fig. 5.5), so that no competition of $\mathrm{Bu_4N^+}$ for the $\mathbf{T}$ hosting cavity is expected. No significant change in the absorption spectrum of a $\mathrm{CH_2Cl_2}$ solution of $\mathbf{D3_2B^{2+}}$ has also been observed upon addition of $\mathrm{NBu_4PF_6}$ (up to 0.14 M). On the other hand, titration of a $\mathrm{CH_2Cl_2}$ solution of $\mathbf{T}$ containing 0.14 M $\mathrm{NBu_4PF_6}$ with $\mathbf{D3_2B^{2+}}$ shows a stronger decrease of fluorescence compared to the same titration performed in the absence of salt (compare the spectra represented by dashed black and gray lines in Fig. 5.5). The association constant raises from 9300 M^{-1} when no salt is present in solution to 23000 M^{-1} at $[\mathrm{NBu_4PF_6}] = 0.14$ M.

Qualitatively similar results have been obtained by titrating with $\mathrm{D3_2B^{2+}}$ dendrimer $\mathrm{CH_2Cl_2}$ solutions of $\mathbf{T}$ containing $\mathrm{NEt_4PF_6}$, $\mathrm{NEt_4BF_4}$, $\mathrm{NBu_4BF_4}$, and $\mathrm{NBu_4BPh_4}$ salts.

Spectrofluorimetric titrations of $\mathrm{CH_2Cl_2}$ air-equilibrated solutions of ca. 10 μM $\mathbf{T}$ at different $\mathrm{NBu_4PF_6}$ concentrations with $\mathbf{D1_2B^{2+}}$, $\mathbf{D3B^{2+}}$, and $\mathbf{D3_2B^{2+}}$ dendrimers and model compound $\mathrm{do}\mathbf{B^{2+}}$ evidence that salt addition affects the thermodynamic stability of the host–guest complexes. In particular, the ratio $K_{\mathrm{ass}}/K_{\mathrm{ass}}^0$, where K_{ass} and K_{ass}^0 are the value of association constants obtained in the presence and absence of salt (Fig. 5.6), respectively, shows that (i) for model compound

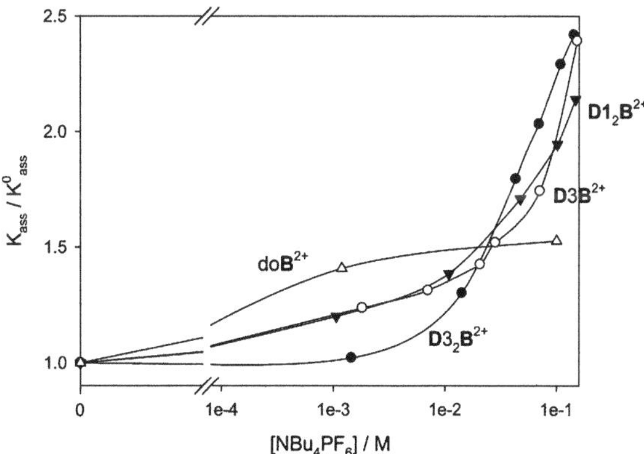

Fig. 5.6 Plot of the association constant ratio obtained in the absence (K_{ass}^0) and in the presence of increasing amounts (K_{ass}) of $\mathrm{NBu_4PF_6}$, for $\mathbf{D3_2B^{2+}}$ (*solid circles*), $\mathbf{D3B^{2+}}$ (*open circles*), $\mathbf{D1_2B^{2+}}$ (*solid triangles*), and $\mathrm{do}\mathbf{B^{2+}}$ (*open triangles*) in air-equilibrated $\mathrm{CH_2Cl_2}$ solution at 298 K

doB^{2+} the effect is small and quantitatively equal upon addition of 1 or 100 mM NBu_4PF_6, (ii) for the investigated dendrimers, upon addition of a high concentration of salt (100 mM) ca. 2.5-fold increase in K_{ass} is observed, but (iii) at low salt concentration (1 mM), the behavior of the three dendrimers is different and K_{ass} is practically unaffected only in the case of $D3_2B^{2+}$.

The observed effect on association constants cannot be attributed to counterion assisted complexation (a few equivalents of salt does not influence complex formation. For references on this topic see [34]) or allosteric effects [35, 36]. Indeed, a large excess of salt is needed (at least 100-fold excess) to see a sizable change in the association constants of dendrimers with T. Moreover, as previously discussed, under the investigated conditions, the bipyridinium core is always ion-paired to its PF_6^- counterions. Therefore, the trend reported in Fig. 5.6 cannot be ascribed to a change in the medium properties: because NBu^{4+} and PF_6^- ions are almost completely associated in CH_2Cl_2 solution, the Debye–Huckel model is not suitable to describe the present effect. An explanation can be obtained from the NMR experiments which show that the addition of NBu_4PF_6 to CH_2Cl_2 dendrimer solutions causes a change in dendrimer conformation consistent with dendron unfolding. Therefore, the observed increase in the association constants demonstrates that dendron unfolding stabilizes the host–guest complex between T and dendrimers. The results also show that the salt concentration required to induce a significant effect on the association constant increases by increasing dendron generation. On the other hand, the modest effect observed for the model compound doB^{2+} can be attributed to the presence of alkyl chains instead of polyaryl-ether dendrons.

Summarizing, an increase in solvent polarity (upon addition of CH_3CN to CH_2Cl_2) leads to a strong decrease of the association constants (more than 1 order of magnitude), while addition of NBu_4PF_6 up to 0.15 M brings about an increase (ca. 2.5-fold) in association constants. These apparently contrasting behaviors can be rationalized considering that in the first case a significant increase of polarity causes a lowering of electron donor–acceptor interactions since both the host and the guest are efficiently stabilized by the solvent. In the latter case, the solvent polarity is only slightly affected by the presence of NBu_4PF_6, which is far from being completely dissociated in CH_2Cl_2 solution, and therefore the observed effect is ascribed to a dendron unfolding process that opens the dendrimer structure so that the resulting host–guest complex is more stable.

5.8 Stopped-Flow Measurements

Measurements of fluorescence intensity of T upon addition of dendrimers in a stopped-flow apparatus enabled information to be gained on the kinetics of host–guest complex formation. The fluorescence intensity at $\lambda > 335$ nm arises from the uncomplexed host molecules and the fitting of the data obtained during the titrations shows that upon addition of dendrimers of the same generation the decay

Table 5.3 Rate constants for the formation (k_f) and dissociation (k_b) of the host–guest complex between tweezer **T** and $\mathbf{D3_2B^{2+}}$ dendrimer in air-equilibrated CH_2Cl_2 solution at 293 K

$\mathbf{T \cdot Dn_mB^{2+}}$	k_f (M^{-1} s^{-1})	k_b (s^{-1})
$\mathbf{D1B^{2+}}$	$\geq 10^8$	$\geq 10^3$
$\mathbf{D2B^{2+}}$	$\geq 10^8$	$\geq 10^3$
$\mathbf{D3B^{2+}}$	$\geq 10^8$	$\geq 10^3$
$\mathbf{D1_2B^{2+}}$	1.9×10^7	700
$\mathbf{D2_2B^{2+}}$	7.7×10^6	430
$\mathbf{D3_2B^{2+}}$	2.8×10^6	300

of the **T** fluorescence is slower in the case of the symmetric than for nonsymmetric ones (Table 5.3). This finding can be accounted for by the two different mechanisms of complexation evidenced by NMR measurements: clipping for symmetric dendrimers and threading for nonsymmetric ones. Moreover, within symmetric $\mathbf{Dn_2B^{2+}}$ dendrimers the clipping process is slower as dendrimer generation increases (Table 5.3), as expected because of the increased steric hindrance of the appended dendrons that have to rearrange their conformation in order to enable the binding of the dendritic core.

The effect of NBu_4PF_6 addition on the kinetics of the complex formation between **T** and $\mathbf{D3_2B^{2+}}$ dendrimer has also been investigated in an air-equilibrated CH_2Cl_2 solution of 14 μM **T** containing 1 equiv of dendrimer at 293 K. The rate of **T** fluorescence decay decreases by increasing the salt content (compare the decays represented by gray squares and black circles in Fig. 5.7). Lower complex formation and dissociation rate constants are obtained upon increasing NBu_4PF_6 concentration (Table 5.4).

This behavior is consistent with the lower diffusion coefficients of the dendrimers obtained from the NMR techniques. A closer inspection to Figs. 5.3 and 5.7

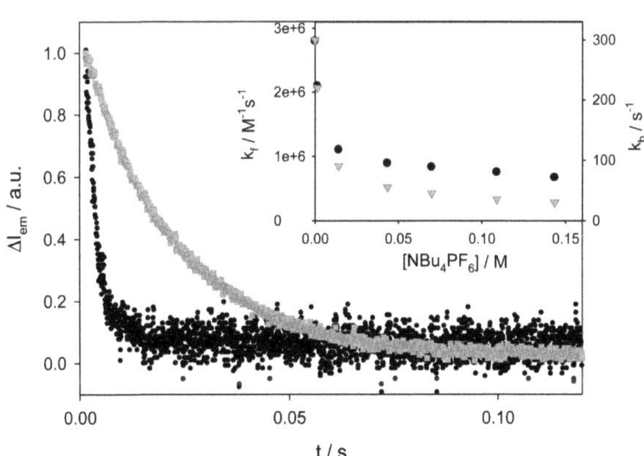

Fig. 5.7 Fluorescence decay of 14 μM **T** air-equilibrated CH_2Cl_2 solution at 293 K containing 1 equiv of $\mathbf{D3_2B^{2+}}$ in the absence (*black circles*) and presence of 0.14 M NBu_4PF_6 (*gray squares*); $\lambda_{ex} = 297$ nm. The *inset* shows the kinetics constants for association (k_f, *black circles*) and dissociation (k_b, *gray triangles*) processes between **T** and $\mathbf{D3_2B^{2+}}$ plotted versus NBu_4PF_6 concentration

Table 5.4 Rate constants for the formation (k_f) and dissociation (k_b) of the host–guest complex between tweezer **T** and **D3$_2$B^{2+}** dendrimer in air-equilibrated CH_2Cl_2 solution at 293 K at different concentrations of NBu_4PF_6

[NBu$_4$PF$_6$] (M)	k_f (M^{-1} s^{-1})	k_b (s^{-1})
0	2.8×10^6	300
1.5×10^{-3}	2.1×10^6	220
1.4×10^{-2}	1.1×10^6	90
4.4×10^{-2}	8.9×10^5	55
7.0×10^{-2}	8.3×10^5	45
1.1×10^{-1}	7.5×10^5	35
1.4×10^{-1}	6.7×10^5	30

shows that the diffusion coefficients of dendrimers decrease linearly with NBu_4PF_6 concentration, while the rate constants k_f and k_b show a stronger dependence on salt concentration. This points to a quite complex kinetic behavior: dendron folding/unfolding processes affect not only the diffusion coefficient of the dendrimers but also the kinetics of formation/dissociation of the host–guest complex.

5.9 Conclusion

Host–guest complex formation between a molecular tweezer **T** and 4,4′-bipyridinium cored dendrimers **Dn_mB^{2+}** has been investigated by NMR, absorption, and emission spectroscopy, as well as electrochemical measurements. The complex formation is mainly driven by electron donor/acceptor interactions. The effect of solvent polarity and addition of NBu_4PF_6 on both the thermodynamic and kinetic features of the association process have been investigated.

As to the thermodynamic properties, in CH_2Cl_2 association constant values (i) are higher for nonsymmetric **DnB^{2+}** dendrimers than for the corresponding symmetric **Dn_2B^{2+}** ones, (ii) within each dendrimer family, they decrease by increasing generation of the dendrons, and (iii) they increase (by ca. 2.5-fold) upon addition of NBu_4PF_6 (0.15 M). In higher polarity mixtures, such as CH_2Cl_2/CH_3CN 1:3 (v/v), association constants are one order of magnitude lower and almost unaffected by dendrimer generation and symmetry. These experimental results can be interpreted as follows: (a) an increase of solvent polarity (by addition of CH_3CN) leads to a better stabilization of the host and guest species by solvent molecules, thus disfavoring charge-transfer interactions between dendrimers and tweezer and (b) addition of NBu_4PF_6 to the solutions in low-polar solvents causes a change in dendrimer conformation that can be ascribed to dendron unfolding. The unfolded structure stabilizes the host–guest complex, as demonstrated by the increase in the association constants.

The kinetic features of complex formation and dissociation show that in a polar solvent the rate constants are higher for the nonsymmetric **DnB^{2+}** dendrimers compared to those for the symmetric **Dn_2B^{2+}** ones, and decrease by increasing dendrimer generation within the symmetric family, and by adding NBu_4PF_6, as a result of dendron unfolding.

The present work demonstrates the importance of taking into account the effects exerted by salt addition when both spectroscopic and electrochemical techniques are exploited to characterize electron donor/acceptor complexes in low polarity solvents, especially when charged species are involved. In such cases, the addition of supporting electrolytes required in electrochemical experiments might modify the dendrimer conformation inducing unfolding of the branches, a behavior that strengthens analogies between dendrimers and proteins [37, 38]. These conformational changes should be considered when designing dendrimers for drug delivery and catalysis purposes or sensing applications.

References

 1. Lehn JM (1995) Supramolecular chemistry: concepts and perspectives. VCH, Weinheim
 2. Reichardt C (2003) Solvents and solvent effects in chemistry, 3rd edn. Wiley-VCH, Weinheim
 3. Newkome GR, Moorefield C, Vogtle F (2001) Dendrimers and dendrons: concepts, syntheses, perspectives. Wiley-VCH, Weinheim
 4. Vogtle F, Richardt G, Werner N (2009) Dendrimer chemistry. Wiley-VCH, Weinheim
 5. Ong W, Gomez-Kaifer M, Kaifer AE (2004) Chem Commun 1677
 6. Sobransingh D, Kaifer AE (2005) Chem Commun 5071
 7. Ong W, Grindstaff J, Sobransingh D, Toba R, Quintela JM, Peinador C, Kaifer AEJ (2005) Am Chem Soc 127:3353
 8. Bruinink CM, Nijhuis CA, Peter M, Dordi B, Crespo-Biel O, Auletta T, Mulder A, Schoenherr H, Vancso GJ, Huskens J, Reinhoudt DN (2005) Chem-Eur J 11:3988
 9. Wang W, Kaifer AE (2006) Angew Chem Int Ed 45:7042
10. Ornelas C, Ruiz Aranzaes J, Cloutet E, Alves S, Astruc D (2007) Angew Chem Int Ed 46:872
11. Hahn U, Cardinali F, Nierengarten JF (2007) New J Chem 31:1128
12. Branchi B, Ceroni P, Balzani V, Bergamini G, Klarner FG, Vogtle F (2009) Chem-Eur J 15:7876
13. Balzani V, Ceroni P, Giansante C, Vicinelli V, Klarner F-G, Verhaelen C, Vogtle F, Hahn U (2005) Angew Chem Int Ed 44:4574
14. Balzani V, Bandmann H, Ceroni P, Giansante C, Hahn U, Klarner F-G, Muller WM, Muller U, Verhaelen C, Vicinelli V, Vogtle FJ (2006) Am Chem Soc 128:637
15. Monk PMS (1998) The viologens: physicochemical properties, synthesis, and applications of the salts of 4, 4′-bipyridine. Wiley, New York
16. Ballardini R, Credi A, Gandolfi MT, Giansante C, Marconi G, Silvi S, Venturi M (2007) Inorg Chim Acta 360:1072
17. Klarner F-G, Burkert U, Kamieth M, Boese MJ (2000) Phys Org Chem 13:604
18. Klarner F-G, Kahlert B (2003) Acc Chem Res 36:919
19. Marchioni F, Juris A, Lobert M, Seelbach UP, Kahlert B, Klarner F-G (2005) New J Chem 29:780
20. Barrire F, Geiger WEJ (2006) Am Chem Soc 128:3980
21. Thompson PA, Simon JDJ (1993) Am Chem Soc 115:5657
22. Masnovi JM, Levine A, Kochi JKJ (1985) Am Chem Soc 107:4365
23. Rosokha SV, Sun D, Fisher J, Kochi JK (2008) ChemPhysChem 9:2406
24. Stewart GM, Fox MAJ (1996) Am Chem Soc 118:4354
25. Ceroni P, Vicinelli V, Maestri M, Balzani V, Muller WM, Muller U, Hahn U, Osswald F, Vogtle F (2001) New J Chem 25:989

26. Amabilino DB, Asakawa M, Ashton PR, Ballardini R, Balzani V, Belohradsky M, Credi A, Higuchi M, Raymo FM, Shimizu T, Stoddart JF, Venturi M, Yase K (1998) New J Chem 22:959
27. Cohen Y, Avram L, Frish L (2005) Angew Chem Int Ed 44:520
28. Johnson CS Jr (1999) Prog Nucl Magn Reson Spectrosc 34:203
29. Stejskal OE, Tanner JEJ (1965) Chem Phys 42:288
30. Macchioni A, Ciancaleoni G, Zuccaccia C, Zuccaccia D (2008) Chem Soc Rev 37:479
31. Denuault G, Mirkin MV, Bard AJJ (1991) Electroanal Chem 308:27
32. Sun H, Chen W, Kaifer AE (2006) Organometallics 25:1828
33. Huang F, Jones J, Slebodnick C, Gibson HWJ (2003) Am Chem Soc 125:14458
34. Jones JW, Zakharov LN, Rheingold AL, Gibson HWJ (2002) Am Chem Soc 124:13378
35. Arduini A, Giorgi G, Pochini A, Secchi A, Ugozzoli FJ (2001) Org Chem 66:8302
36. Kubik SJ (1999) Am Chem Soc 121:5846
37. Beringhelli T, Eberini I, Galliano M, Pedoto A, Perduca M, Sportiello A, Fontana E, Monaco HL, Gianazza E (2002) Biochemistry 41:15415
38. Jelesarov I, Eberhard D, Thomas RM, Bosshard HR (1998) Biochemistry 37:7539

Chapter 6
Self-Assembly of Calix[6]arene-Diazapyrenium Pseudorotaxanes: Interplay of Molecular Recognition and Ion-Pairing Effects

6.1 Introduction

Pseudorotaxanes are supramolecular systems minimally composed of an axle-like molecule surrounded by a macrocyclic component [1]. These peculiar host–guest systems represent convenient precursors for the synthesis of more complex inter-locked structures, such as rotaxanes and catenanes [2–13], which can operate as simple molecular machines. Essentially, the functioning mode of a pseudorotaxane consists of an external stimulus-promoted reversible threading–dethreading process. The structural information stored in both axle and wheel should match (be complementary) from a dimensional point of view as well as for the inter-component interactions responsible for the stability of the complex. A common feature of these systems is that the macrocycle resides in a precise position (station) along the axle, which is also the structural unit that responds to the external stimuli governing the dethreading–threading process. Typically this unit is positioned in a more or less central portion of the axial component. In a rotaxane, the two stoppers present at the termini of the axial component, by acting as a structural tie, prevent dethreading.

The ability of these systems to perform functions relies on their rational design, which in turn requires a deep knowledge of the interactions that occur among the molecular components. As the inter-component interactions that drive the recognition are electrostatic in nature, low-polarity solvents are frequently employed to maximize the efficiency of the self-assembly process. In many types of pseudorotaxanes, however, one of the components (usually the axle) is a charged species that is subject to ion pairing in apolar media [14–20]. Such ion-pairing equilibria can strongly affect the threading reaction, thereby complicating the rationalization of the self-assembly process. In the case of a charged axle, these effects can be circumvented or even advantageously exploited to improve the efficiency of the host–guest recognition process, by utilizing a heteroditopic macrocyclic component, which can simultaneously bind the guest and its counterions [21–31].

M. Baroncini, *Design, Synthesis and Characterization of new Supramolecular Architectures*, Springer Theses, DOI: 10.1007/978-3-642-19285-2_6,
© Springer-Verlag Berlin Heidelberg 2011

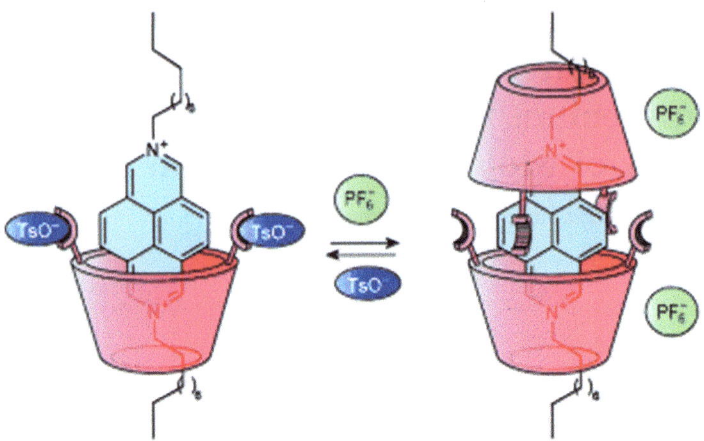

We have recently studied the ability of the tris(phenylureido)calix[6]arene **CX** to act as a heteroditopic and non-symmetrical wheel to form oriented pseudorotaxanes with non-symmetrical 1,1-dialkyl-4,4′-bipyridinium axles [32, 33]. This three-dimensional macrocycle possesses a π-electron-donor cavity that can interact with electron-accepting positively charged units, and three efficient hydrogen-bonding donor phenylureido groups at the upper rim that, besides extending the cavity, can complex the counterions of the cationic axle, thus assisting its insertion into the wheel from this rim. By exploiting these peculiar binding properties, rotaxanes characterized by the unidirectional orientation of the two stoppers of the dumbbell with respect to the two calixarene rims were realized [34, 35] (Chart 6.1).

Along this line, we envisaged that the electron-accepting dication 2,7-diaza-pyrenium (DAP) unit present in the N,N-didecyl-2,7-diazapyrenium axle-like molecule $\mathbf{1} \cdot 2\mathrm{PF}_6$ (proton labelling as shown), because of its large size, could serve both as stopper and station (act as an active stopper) for the construction of

Chart 6.1 Structural formulas of the wheel **CX** and N,N'-didecyl-2,7-diazapyrenium thread-like molecule $\mathbf{1} \cdot 2\mathrm{PF}_6$

pseudorotaxanes that can thread and dethread only from one side of the wheel. In addition, because of its spectroscopic features, it could also signal its proximity to a specific rim of the calixarene wheel. In fact, the DAP unit of compound 1^{2+}, like the previously studied 4,4′-bipyridinium moiety, is a cationic species that has already been used as the guest in supramolecular inclusion complexes [36–40]; 1^{2+} has an extended π delocalization that confers characteristic spectroscopic features to this chromophore [41]. Its sharp, intense and structured absorption and luminescence bands can be easily monitored to reveal any alteration in the surrounding environment. The aim of this work is to study the self-assembly process of compounds 1^{2+} and **CX**, and to investigate the parameters that affect pseudorotaxane formation.

6.2 NMR Spectroscopic Characterization

NMR spectroscopic characterization: The 1H NMR spectrum of $1\cdot2PF_6$ taken in CD3CN (see Fig. 6.1a) shows a singlet at $\delta = 9.88$ ppm for the four aromatic protons and a singlet at $\delta = 8.85$ ppm for the four protons. The two equivalent methylene protons 10 resonate as a triplet at $\delta = 5.09$ ppm, whereas all other

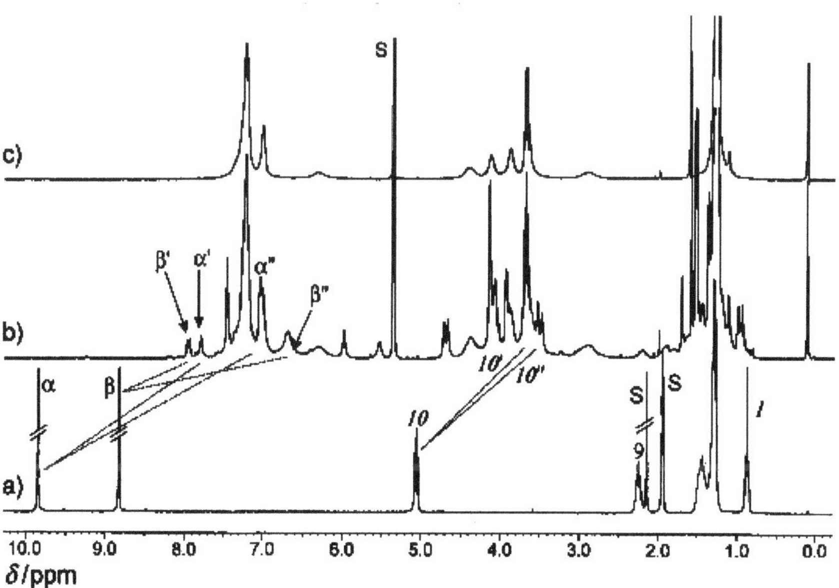

Fig. 6.1 1H NMR stack plot (300 MHz) of: **a** $1\cdot2PF6$ in CD3CN; **b** a mixture of **CX** and $1\cdot2PF6$ in CD_2Cl_2 (the *dotted lines* indicate a few representative resonances of 12+ that are up-field shifted upon complexation) and **c CX** in CD_2Cl_2

protons of the two alkyl chains resonate in the 2.3–0.8 ppm range. $\mathbf{1}\cdot\mathbf{2PF_6}$ is not sufficiently soluble in low-polarity media for accurate NMR analysis; however, by adding a slight excess of the salt to solution of $\mathbf{CX}$ in CDCl3, C6D6 or CD_2Cl_2, the appearance of a reddish colour of the solution suggests that in all instances a binding process between $\mathbf{1}\cdot\mathbf{2PF_6}$ and $\mathbf{CX}$ has taken place. In CD_2Cl_2 the 1H NMR spectrum showed sharper signals than those obtained in the other solvents, so it was selected to study the structure of the species present in solution and the stoichiometry of binding. The spectrum of the sample obtained after filtration of a 10 mM solution of $\mathbf{CX}$ containing an equivalent amount of $\mathbf{1}\cdot\mathbf{2PF_6}$ clearly shows the presence of uncomplexed $\mathbf{CX}$ (see Fig. 6.1c for the spectrum of free $\mathbf{CX}$) together with a 1:1 adduct $\mathbf{CX} \supset \mathbf{1}\cdot\mathbf{2PF_6}$ (see Fig. 6.1b).

Upon binding, $\mathbf{CX}$ adopts a pseudo-cone conformation as shown, for example, by the presence of an AX system at $\delta = 4.70$ and 3.51 ppm for the six axial and six equatorial methylene protons of the bridges, respectively. The methoxy groups, which in the free wheel resonate as a very broad signal at 3 ppm, in the complex are visible as a downfield-shifted sharp singlet at 4.14 ppm, thus suggesting their guest-induced expulsion from the interior of the cavity [32, 33, 34, 35], The binding also appreciably affects the resonances of the guest protons as seen, for example, by the large up-field shift experienced by the signals of protons α and β belonging to the DAP unit. The former of these two signals splits into two singlets at $\delta = 7.79$ (α') and 7.05 ppm (α''), whereas the latter splits into two doublets at $\delta = 7.95$ (β') and 6.7 ppm (β'') with a coupling constant of 9 Hz (see Fig. 6.1b). The signals of protons α'' and β'' were overlapped by more intense signals arising from the aromatic protons of $\mathbf{CX}$. Their full assignment was thus accomplished through 2D COSY and HSQC experiments. The methylene groups 10 give rise to two very broad signals at $\delta = 3.7$ (10′) and 3.6 ppm (10″). Both signals are up-field shifted ($\Delta\delta$ up to 1.5 ppm) with respect to the free $\mathbf{1}\cdot\mathbf{2PF_6}$ (cf. Fig. 6.1a, b) and are fully overlapped by stronger resonances arising from the 2-ethoxy-ethoxy chains present on the calix[6]arene lower rim. Nevertheless, their nature was verified through 2D COSY and TOCSY experiments.

The analysis of the integrals of the spectrum clearly supports the formation of an adduct with a 1:1 stoichiometry as the main host–guest complex. However, the variations of chemical shift observed for both $\mathbf{CX}$ and $\mathbf{1}^{2+}$ could be consistent with different structures of the complex. Namely, a pseudorotaxane structure in which one of the two alkyl chains of $\mathbf{1}^{2+}$ protrudes from the lower rim of the wheel (see Scheme 6.1a) or one in which the DAP unit of $\mathbf{1}^{2+}$ seats orthogonally to the upper rim of the calix[6]arene (Scheme 6.1b). The large up-field shift experienced by the resonances of α'' ($\Delta\delta$ 2.8 ppm) and β'' ($\Delta\delta$ 2.1 ppm) protons supports the hypothesis that a pseudorotaxane complex has formed in solution. In this structural arrangement, indeed, one part of the DAP unit is engulfed in the aromatic cavity of $\mathbf{CX}$, thus exposing its protons to a more intense anisotropic shielding effect.

The small chemical shift differentiation endured by 10′ and 10″ could be explained by considering that, differently from what was previously observed for 1,1-dialkyl-4,4′-bipyridinium-based axles, the DAP unit of $\mathbf{1}^{2+}$, because of its size,

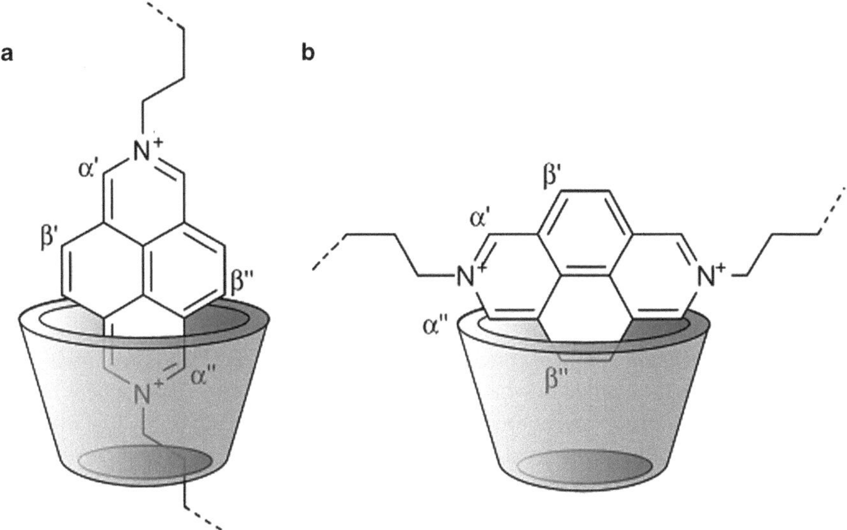

Scheme 6.1 Schematic representation of the two possible structures of the complex formed between the axle **1**·2PF$_6$ and wheel **CX**

is less engulfed in the aromatic cavity of the macrocycle and cannot protrude from its lower rim. This could impose on the two methylene groups 10′ and 10″ a similar anisotropic shielding effect, exerted by the calix[6]arene cavity and the phenyl groups present on the upper rim of **CX**. The threading reaction also accounts for the coupling observed between protons β' and β'', which are no longer magnetically equivalent. 2D ROESY experiments of the complex finally confirmed the hypothesis, since there are several NOE cross-peaks arising from the spatial proximity between, for example, the methoxy groups present on the lower rim of **CX** and the methylene groups belonging to only one alkyl chain of **1^{2+}**, and also between the β' protons of DAP and the tert-butyl groups present on the upper rim of the macrocycle.

A few weak signals present in the spectrum of the 1:1 axle-wheel mixture of Fig. 6.1b could not be assigned either to the pseudorotaxane adduct **CX ⊃ 1**·2PF$_6$ or to the free **CX**. We formulated the hypothesis that these signals could arise from the formation of axle-wheel adducts having a stoichiometry higher than 1:1. To disclose the nature of all the species present in dichloromethane solution, some diffusion-ordered spectroscopy (DOSY) experiments were carried out ([42], for a general review on the application of DOSY in supramolecular chemistry, please see: [43, 44]). A mean diffusion coefficient D of $(0.68 \pm 0.02) \times 10^{-5}$ cm^2 s^{-1} for **CX** was calculated by performing the NMR diffusion experiment on a 10 mM solution of the free **CX** in CD$_2$Cl$_2$. The experiment was then repeated under the same experimental conditions on the mixture containing **CX ⊃ 1**·2PF$_6$ and yielded values of D that, depending on the resonance selected, ranged between 0.55

and 0.65×10^{-5} cm^2 s^{-1}. The highest values were assigned to the resonances of the free **CX**, whereas the attenuation profiles of all resonances previously assigned to the pseudorotaxane, such as β' and α' (see Fig. 6.2), afforded diffusion coefficients ranging from 0.56 to 0.58×10^{-5} cm^2 s^{-1}.

The simple relationship $\mathrm{MW}_{(1)}/\mathrm{MW}_{(2)} = (D_{(2)}/D_{(1)})$ derived from the Stokes equation correlates the molecular weight (MW) and the diffusion coefficient of two spherical species, (1) and (2), diffusing in solution under the same experimental conditions. The application of this equation to the diffusion data measured from the resonances of the pseudorotaxane yielded, despite the non-spherical shape of the diffusing species, a molecular weight range between 2360 and 2620 Da, which is in good agreement with the theoretical value of 2241 Da (the reduction of the magnitude of the diffusion coefficient measured upon binding ($\approx 17\%$) is comparable to that found in similar studies in which an α-cyclodextrin was used as a wheel for the formation of pseudorotaxane compounds, see: [45]). Very interestingly, the fitting of the attenuation profile of the weak doublet at $\delta = 8.2$ ppm (, see Fig. 6.2a) yielded an unexpectedly low value of D (($0.47 \pm 0.04) \times 10^{-5}$ cm^2 s^{-1}). This value, though inaccurate, was an indication that in dichloromethane solution the formation of adducts having a stoichiometry higher than 1:1 is possible.

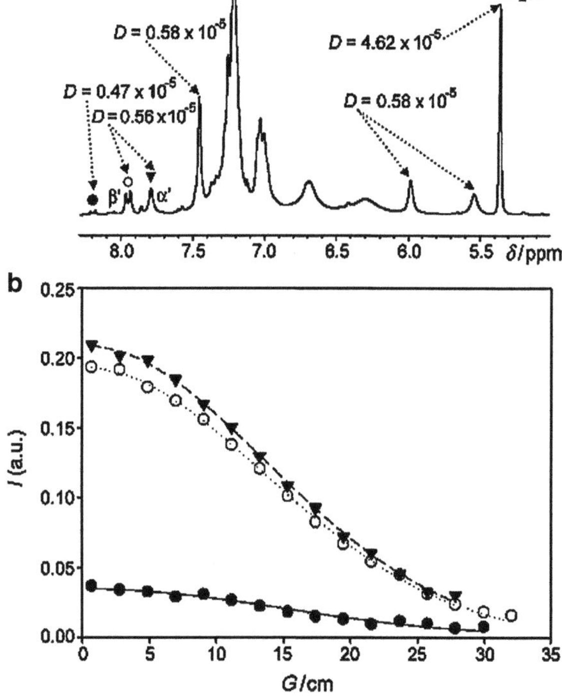

Fig. 6.2 a Expansion of the ^{1}H NMR spectrum (300 MHz) of a mixture containing the pseudorotaxane complex **CX** $\supset$ **1**·2PF$_6$. The diffusion coefficients D calculated through NMR diffusion experiments are reported above the most representative peaks. **b** Plot of the intensity attenuation profiles of resonances *filled circle, open circle, filled inverted triangle,* and upon gradient variation

6.3   ESIMS Experiments

To determine the nature of these higher-stoichiometry adducts, the previous mixture was submitted to high-resolution ESIMS analysis. The spectrum recorded in the 700–2000 mass range shows the presence of a peak at m/z 1709.01 for the doubly charged 2:1 adduct $[(\mathbf{CX})_2 \supset \mathbf{1}]^{2+}$, along with the wheel $\mathbf{CX}$ (visible as the adduct with Na$^+$) and the most abundant doubly charged 1:1 adduct $[\mathbf{CX} \supset \mathbf{1}]^{2+}$ at m/z 976.10. Singly charged adducts that should retain at least one PF$_6{}^-$ anion were never observed.

6.4   UV/Vis Spectroscopic Experiments

All photophysical experiments were performed in air-equilibrated CH$_2$Cl$_2$ solution at room temperature. The absorption and fluorescence spectra of compound $\mathbf{1}^{2+}$ show the typical [41] structured bands of the DAP chromophoric unit. The calixarene wheel $\mathbf{CX}$ shows an intense absorption band in the UV region of the spectrum and a weak luminescence (Table 6.1).

To investigate the complexation equilibrium between $\mathbf{1}\cdot 2\mathrm{PF}_6$ and $\mathbf{CX}$, we performed UV/Vis titration experiments. The addition of increasing amounts of the wheel component to the DAP-based thread caused remarkable changes in both the absorption (Fig. 6.3) and emission spectra of $\mathbf{1}\cdot 2\mathrm{PF}_6$ [experiment (i)], namely, the absorption bands of the transition to the first $\pi-\pi^*$ excited state ($350 < \lambda < 450$ nm) shifted to longer wavelengths, the bands corresponding to the transition to the second $\pi-\pi^*$ excited state ($300 < \lambda < 350$ nm) broadened and weakened, and a broad absorption tail at $\lambda > 450$ nm appeared; the luminescence of the DAP moiety was quenched.

The absorption and luminescence data were fitted with the SPECFIT program [46], which optimizes the interpolation of the titration data according to a given complexation model utilizing the whole investigated spectral range. The best fit was obtained by assuming the formation of both the 1:1 and 2:1 host–guest complexes [Table 6.2, experiment (i)]. To clarify the titration results, we also constructed several Job plots [47–49], by performing the experiments at different concentrations and reading the absorption changes at various wavelengths. The Job plots show that the molar fraction corresponding to the maximum (or minimum) of the plots depends on the concentration and the wavelength; this result

Table 6.1 Absorption and emission data for $\mathbf{CX}$ and $\mathbf{1}\cdot 2\mathrm{PF}_6$ (CH$_2$Cl$_2$, 293 K)

Compound	λ_{max} (nm)	$\varepsilon \times 10^{-3}$ (m^{-1} cm^{-1})	λ_{em} (nm)	ϕ	τ (ns)
$\mathbf{1}\cdot 2\mathrm{PF}_6$	258	58	427	0.39	7.7
	338	35			
CX	421	15	330	0.003	< 1
	250	54			

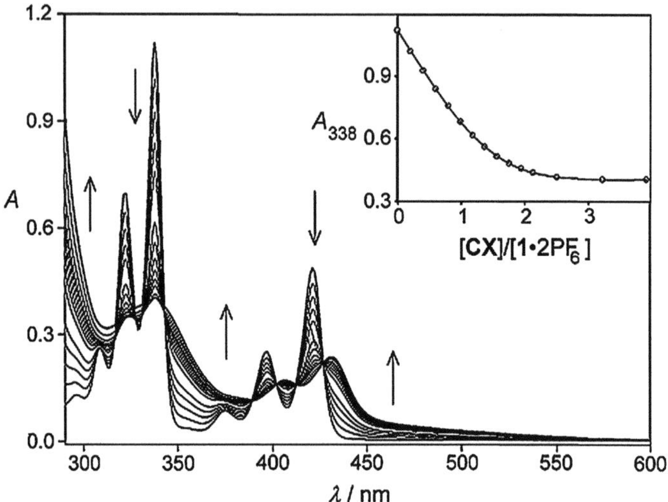

Fig. 6.3 Absorption spectra of compound **1·2PF$_6$** (3.3×10^{-5} M) upon titration with **CX**, CH$_2$Cl$_2$, RT. The *inset* shows the absorption changes at 338 nm (*open circle*) and the fitting curve (-) obtained by means of the SPECFIT fitting program

Table 6.2 Thermodynamic and kinetic constants (CH$_2$Cl$_2$, RT)

Expt.	$K_1 \times 10^{-5}$ (M^{-1})[a]	$K_2 \times 10^{-5}$ (M^{-1})[b]	$k_1 \times 10^{-2}$ (M^{-1} s^{-1})[c]	$k_1 \times 10^{3}$ (s^{-1})[c]	$k^2 \times 10^{-2}$ (M^{-1} s^{-1})[c]
(i)[d]	3 ± 0.5	1.9 ± 0.3	1.0 ± 0.2	2.8 ± 0.6	1.6 ± 0.3
(ii)[e]	0.50 ± 0.02				
(iii)[f]	1000 ± 300				

[a] Observed stability constant referred to Eq. 1, Scheme 6.2
[b] Observed stability constant referred to Eq. 2, Scheme 6.2
[c] Rate constants referred to Eqs. 1 and 2, Scheme 6.2; values calculated by means of the SPECFIT software
[d] Titration of **1·2PF$_6$** with **CX**
[e] Titration of a solution of **1^{2+}** containing an excess of PF$_6^-$ anions with **CX**
[f] Titration of **1^{2+}** with a solution of **CX** containing two equivalents of TsO$^-$ anions

demonstrates that there is more than one type of adduct. Even though these experiments do not fully elucidate the stoichiometry of the complexes, they support the interpretation of the UV/Vis spectroscopic titration.

To confirm the stoichiometry of the host–guest complexes and gain more information on the mechanism of association, we performed kinetic experiments by mixing CH$_2$Cl$_2$ solutions of **CX** and **1·2PF$_6$** in 0.7:1 and 2:1 molar ratios, to favour the formation of the 1:1 and 2:1 complexes, respectively. The kinetic traces recorded by monitoring absorption changes at 421 nm (see Fig. 6.4) and 338 nm were fitted with the SPECFIT fitting program. The results show that, upon mixing **CX** and **1·2PF$_6$** in 0.7:1 molar ratio, the 1:1 complex is formed in two steps,

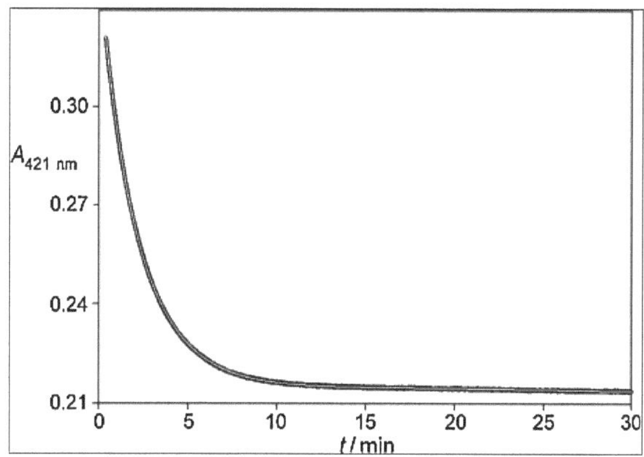

Fig. 6.4 Kinetic trace (*black line*), recorded in CH_2Cl_2 at 293 K, for the absorbance changes at 421 nm obtained upon mixing **CX** (5.3×10^{-5} M) and **1·2PF₆** (2.5×10^{-5} M). The white line is the fitting curve obtained by means of the SPECFIT fitting program

Scheme 6.2 Kinetic models used to fit the kinetic traces. The counterions are not reported for reasons of clarity

$$\mathbf{1 + CX} \underset{k_{-1}}{\overset{k_1}{\rightleftharpoons}} \mathbf{[CX \supset 1]^{\#}} \overset{k_1'}{\longrightarrow} \mathbf{[CX \supset 1]} \qquad (1)$$

$$\mathbf{[CX \supset 1] + CX} \underset{k_{-2}}{\overset{k_2}{\rightleftharpoons}} \mathbf{[(CX)_2 \supset 1]} \qquad (2)$$

namely a second-order association reaction followed by a first-order rearrangement of the adduct (Eq. 1 in Scheme 6.2).

Upon mixing **CX** and **1·2PF₆** in 2:1 molar ratio, the best fits were obtained by assuming the reaction sequence illustrated by Eqs. 1 and 2 in Scheme 6.2. First, the 1:1 adduct is formed in two steps (Eq. 1), with the same kinetic rate constants determined in the previous experiment. Successively, this complex reacts with a second calixarene molecule, forming the 2:1 complex with a second-order reaction (Eq. 2). The values of the kinetic rate constants are gathered in Table 6.2.

From these results we can conclude that under the experimental conditions of the UV/Vis spectroscopic measurements **CX** and **1²⁺** can also form a 2:1 host–guest complex, which is not observed in the NMR experiments. The main difference between these measurements is the concentration of the molecular components: the concentration of the examined compounds in the UV/Vis spectroscopic experiments ($[\mathbf{1·2PF_6}] < 3.5 \times 10^{-5}$ M) was two orders of magnitude lower than that in the NMR measurements. The concentrations that can be used in the UV/Vis spectroscopic experiments are determined by the absorption coefficient of the chromophores, the association constant of the adduct and the low solubility of **1·2PF₆** in CH_2Cl_2. Because compound **1·2PF₆** is a salt, its concentration affects the position of the equilibrium between the tight ion pairs and the dissociated ions in solution, and therefore the relative abundance of different charged species. Moreover, the calixarene wheel is a heteroditopic receptor, which can host cations inside its cavity and complex anions by means of the urea moieties on the upper rim [50, 51]. Therefore, we tried to mimic the experimental conditions of the NMR measurements by changing the concentration and the nature of the counteranions of compound 1^{2+}. We performed UV/Vis complexation experiments under two paradigmatic conditions: the titration of a solution of 1^{2+}, which contained an excess of hexafluorophosphate anions, with **CX** [experiment (ii)], and the titration of 1^{2+} with a solution of **CX** containing two equivalents of tosylate (TsO$^-$) anions [experiment (iii)]. PF$_6^-$ and TsO$^-$ anions were selected as poorly and strongly H-bonding species, respectively; moreover, TsO$^-$ anions also possess an aromatic moiety, which could be involved in π-stacking interactions with both **CX** and 1^{2+}.

Before performing experiments (ii) and (iii), we investigated the effect of the concentration and nature of the anions on the photophysical properties of the DAP-based axle. The absorption and luminescence spectra of compound 1^{2+} did not change in presence of 100 equivalents of tetrabutylammonium hexafluorophosphate (TBAPF$_6$) salt. Despite the lack of spectroscopic evidence, it is likely that 1^{2+} and PF$_6^-$ ions are in equilibrium with the ion-paired salt, and that the excess of PF$_6^-$ ions displaces this equilibrium towards the associated ion pair ((a) it is known that in the case of 4,4′-bipyridinium derivatives the addition of an excess of PF$_6^-$ ions to the TsO$^-$ salt causes the exchange of the counterions, see: Ref. [50]; (b) in this case we could not perform the experiment on 1·2TsO because of precipitation). On the other hand, the addition of two equivalents of tetraethylammonium tosylate (TEATsO) caused remarkable changes in both the absorption and emission spectra of 1^{2+} (see Fig. 6.5), comparable to the changes observed upon titration of 1^{2+} with **CX** (see Fig. 6.3). TsO$^-$ ions are more coordinating than PF$_6^-$ ions and can be involved in π-stacking interactions; hence the addition of TEATsO to 1·2PF$_6$ led to the exchange of the counterions of the DAP cation, with the formation of the tight ion pair 1·2TsO. Moreover, soon after the addition of the salt a precipitate started to form, thus preventing any quantitative consideration, but proving that the ion-pairing interactions between 1^{2+} and TsO$^-$ are quite strong. Despite the precipitation of the salt, we noted that the first equivalent of TsO$^-$ ions is responsible for most of the spectroscopic changes. Upon addition of

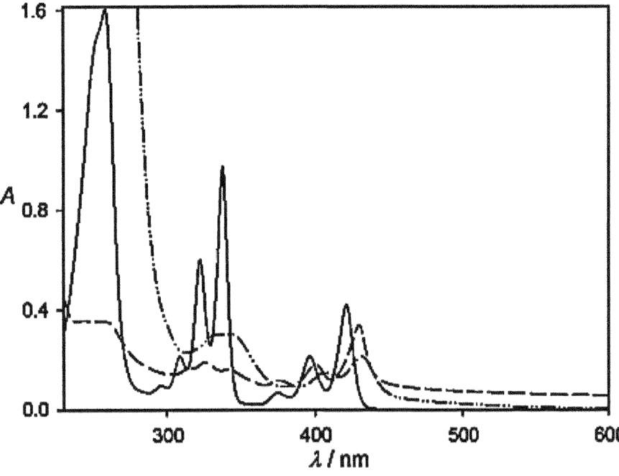

Fig. 6.5 Absorption spectra of compound 1·2PF$_6$, 3.3×10^{-5} M (-), upon successive addition of two equivalents of TEATsO (- - -) and two equivalents of **CX**, after 2 days (······), CH$_2$Cl$_2$, RT. Note that addition of TEATsO to 1·2PF$_6$ causes precipitation of the salt (see text for details)

two equivalents of **CX** to this solution, we observed a progressive solubilization of the precipitate, most likely as a consequence of the formation of the inclusion complex; after 2 days the solution was clear. From a qualitative comparison of the spectra of **1·2TsO** free and complexed with **CX** (see Fig. 6.5), we observed a further shift towards lower energies of the bands at $\lambda > 350$ nm and an increase and loss of structure of the bands in the range 300–350 nm as a consequence of the inclusion of the axle inside the cavity of the calixarene.

As far as experiments (ii) and (iii) are concerned, we performed the absorption and fluorescence titrations of a solution of **1·2PF$_6$** containing 100 equivalents of TEAPF$_6$ with **CX**, and of **1·2PF$_6$** with a solution of **CX** containing two equivalents of TEATsO. The spectroscopic changes resemble those obtained in the titration of **1·2PF$_6$** with **CX** (see Fig. 6.3). The absorption and luminescence data were best fitted with the SPECFIT program assuming uniquely the formation of a 1:1 host–guest complex. The association constants are gathered in Table 6.2.

Taken together, the results of experiments (i)–(iii) clearly show that the concentration and nature of the anions affect both the stoichiometry and the stability of the supramolecular adducts formed by **1^{2+}** and **CX**. A possible explanation can be

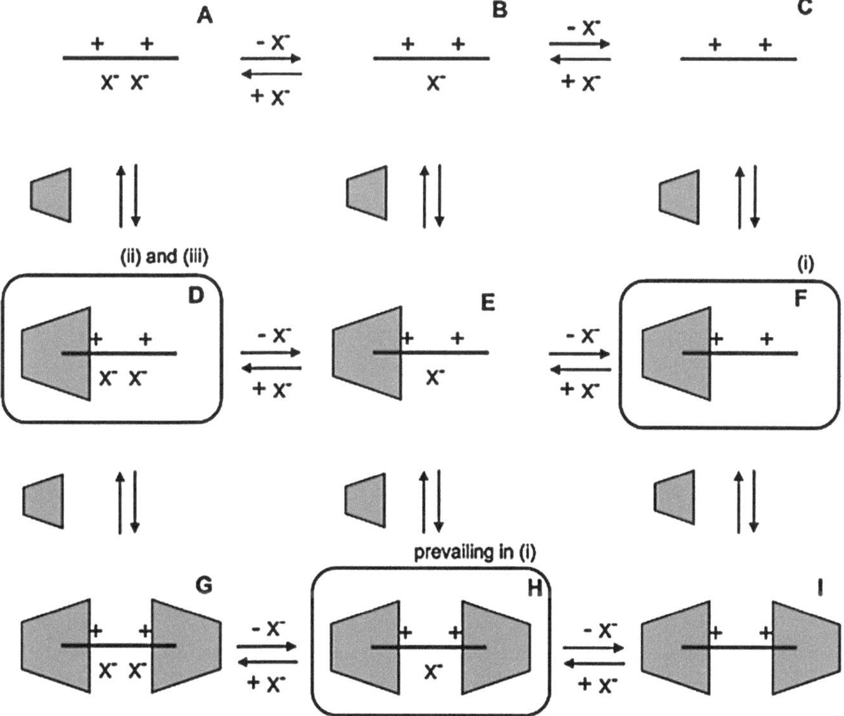

Scheme 6.3 Equilibria for ion-pair dissociation (*horizontal processes*) and host–guest association (*vertical processes*) of the investigated system. Representation of the species in G-I as a threaded complex is hypothetical (see the text for more details)

found by considering that in this system several equilibria can exist, as summarized in Scheme 6.3. The horizontal reactions represent the ion-pairing equilibria, which involve the free axle, as well as the 1:1 and 2:1 host–guest complexes; because the calixarene is also an anion receptor, these equilibria are obviously influenced by the hydrogen-bonding interactions between the anions and the urea moieties of the wheel. The vertical reactions represent the association equilibria, which involve the axle in its differently ion-paired forms. A quantitative, exhaustive investigation of all these equilibria is not possible with our data, but some reasonable comments can be made.

First of all, the portion of the axle that penetrates the cavity of the calixarene has to be free from its counterions for steric reasons [50]. Molecular models and the results obtained for related complexes [50] indicate that the species involved in the formation of the adducts are most likely the free and one-anion-paired axles (B and C in Scheme 6.3). It follows that the calculated association constants are apparent values, because they do not take into account the ion-pairing equilibria. The observed stability constants in experiments (ii) and (iii) are one order of magnitude lower and three orders of magnitude larger, respectively, than in experiment (i) (Table 6.2). As already pointed out, the anions play a dual role. On the one hand, they can interact with the axle, displacing the ion-pairing equilibria towards the associated salt; as a consequence, the active monocationic and dicationic species (B and C in Scheme 6.3) are subtracted from the complexation equilibria (vertical processes in Scheme 6.3) and the observed association constant is lower than the real one. On the other hand, the anions can interact with the diphenylurea units on the upper rim of the wheel, thus concurring in stabilizing the host–guest complex. In experiment (ii), a solution of $1 \cdot 2PF_6$ containing 100 equivalents of $TEAPF_6$ was titrated with **CX**; the main effect of these poorly coordinating anions [51] is to displace the ion-pairing equilibria towards the associated salt, and thus the observed stability constant decreases with respect to experiment (i) (Table 6.2). In experiment (iii), a solution of $1 \cdot 2PF_6$ was titrated with a solution of **CX** containing two equivalents of TEATsO; TsO^- anions are strongly coordinating [51], and can interact with both the axle (see above) and the diphenylurea moieties of the wheel through hydrogen-bonding and -stacking interactions. Under these experimental conditions the wheel, by interacting with TsO^-, seems to be a better host for the axle, at least for two reasons: it bears in the upper rim the TsO^- ions which pre-organize the wheel and exhibit a high coordinating ability towards the axle; moreover, it becomes a negative species. Indeed, the observed stability constant in experiment (iii) is three orders of magnitude larger than in experiment (i).

We can therefore identify three main energetic factors that contribute to the stabilization of the supramolecular complexes: (a) the interaction between the axle and the cavity of the wheel, (b) the hydrogen-bonding interactions between the urea moieties on the upper rim of the calixarene and the anions, and (c) the electrostatic interactions between the cation and the anions held together in the supramolecular structure. In the absence of other effects and if these interactions were additive, it would be expected that the most stable species is $[(\mathbf{CX})_2 \supset \mathbf{1}] \cdot 2X$ (G in

Scheme 6.3), in which the axle is surrounded by two macrocycles and two anions. We have no evidence of the formation of this complex, at least under the majority of the investigated experimental conditions (experiments (ii) and (iii) and NMR experiments).

To explain this behaviour we have to admit that there is competition between the calixarene and the anions for the interaction with the cation, in which steric effects play an important role. From these considerations and a detailed analysis of the spectra, we can infer some information on the identity of the species obtained under different experimental conditions. First of all, to evaluate the extent of the interaction between the axle and the cavity of the wheel, an analysis of the absorption spectra in the region $240 < \lambda < 300$ nm can be very useful. In this region the calixarene makes a substantial contribution to the absorption spectrum, and the encapsulation of a cationic guest inside the wheel considerably affects the absorption spectrum of the calixarene. Therefore, we compared the sum of the spectra of $\mathbf{1 \cdot 2PF_6}$ and $\mathbf{CX}$ in a 1:2 ratio with the spectrum of a mixture of the molecular components in the same proportion. This comparison revealed that the two spectra are very similar in the region at $\lambda < 300$ nm: such an observation suggests that the cationic portion of the axle is scarcely encapsulated inside the cavity of the wheel, presumably because of the relatively large size of the DAP unit.

In previously studied calixarene systems [52] with a bipyridinium-type axle, the encapsulation is more efficient and there is no evidence of formation of 2:1 host–guest complexes. On the other hand, a bipyridinium unit can indeed form a 2:1

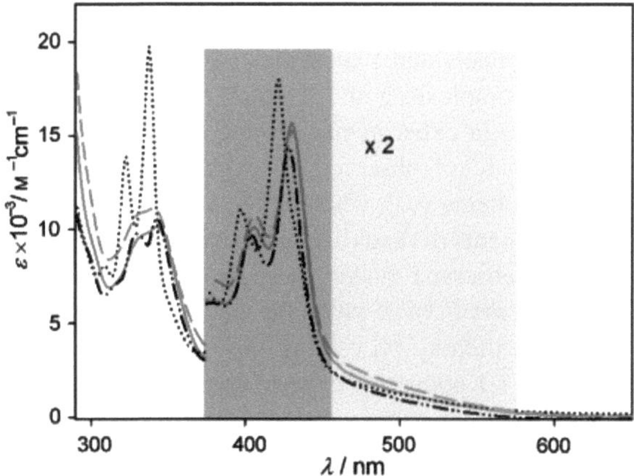

Fig. 6.6 Absorption spectra of the host–guest complexes in CH$_2$Cl$_2$, RT: 1:1 (…, *black line*) and 2:1 (- - -, *grey line*) adducts from experiment (i), 1:1 complexes from experiments (ii) (-·····, *black line*) and (iii) (-, *grey line*). The *shaded areas* reveal the effects on the absorption spectra that we ascribed to the charge-transfer interaction between the axle and the wheel (*light grey*) and to the interaction between the anions and the cationic portion of the axle held together in the complex (*dark grey*)

host–guest complex with a calix[4]arene, the small cavity of which affords a less efficient encapsulation. Additionally, from a comparison of the absorption spectra of the adducts in the three experiments (see Fig. 6.6), we can guess that the 1:1 complex obtained in experiment (i) is different from the 1:1 complex of experiments (ii) and (iii). Both complexes show the absorption tail at $\lambda > 450$ nm, which is ascribed to a charge-transfer interaction between the electron-accepting DAP unit and the electron-rich diphenylurea moieties on the calixarene. Moreover, the spectra obtained in experiments (ii) and (iii) display a shift of the bands in the range 350–450 nm, which can tentatively be assigned to the interaction between the charged axle and the anions held together in the supramolecular structure, as suggested by the shape of the spectrum of the tight ion pair $1\cdot2\text{TsO}$ (see Fig. 6.6). These differences would suggest that under the dilute conditions of experiment (i) the interaction with the anions is lower than that in experiments (ii) and (iii), and that these adducts differ in the extent of ion pairing between the axle and the anions.

Upon addition of **CX** to the 1:1 adducts, only under the conditions of experiment (i) do we observe the formation of the 2:1 host–guest complex. The absorption spectrum of this species (see Fig. 6.6) shows an increase of the charge-transfer band and a shift towards lower energies of the bands of the transition to the first π–π^* excited state ($350 \ll 450$ nm). These observations would suggest, respectively, that the DAP unit is involved in charge-transfer interactions with the aromatic units of both wheels, and that at least one anion is engaged in the supramolecular complex. Although we do not have direct structural evidence in support of a pseudorotaxane topology for the 2:1 adduct (any attempt to obtain crystals of the 2:1 adduct suitable for X-ray analysis failed), alternative structures (e.g., the 1^{2+} guest sitting orthogonally to the rims of two wheels without penetrating the respective cavities, or interacting with the calixarene walls in an alongside fashion) do not fully account for the observed spectra. The 2:1 adduct cannot interact with two anions, because this would imply that two wheels and two anions can coexist in the supramolecular complex, and therefore complexes with 2:1 stoichiometry (G in Scheme 6.3) would be expected to form also in experiments (ii) and (iii). From this analysis we can conclude that the 2:1 adduct observed in experiment (i) is the monocationic $[(\mathbf{CX})_2 \supset \mathbf{1}]\cdot\text{X}^+$ (H in Scheme 6.3), whereby the axle is surrounded by two wheels and one anion. In experiments (ii) and (iii) this species is not formed, so we conclude that the 1:1 complex observed under these conditions is the neutral species $[\mathbf{CX} \supset \mathbf{1}]\cdot2\text{X}$ (D in Scheme 6.3), in which the axle is surrounded by one wheel and two anions are present, thereby preventing the coordination of a second wheel molecule. As far as the 1:1 adduct of experiment (i) is concerned, the absorption spectrum would suggest that this is the dicationic species $[\mathbf{CX} \supset \mathbf{1}]^{2+}$ (F in Scheme 6.3), the axle of which is surrounded just by one wheel, and it is therefore able to take up a second calixarene.

In summary, the self-assembly of **CX** and the DAP-based axle is a complex phenomenon that involves three types of species (wheel, axle and anions) that can mutually interact. When the species come together, these interactions affect one another because of steric effects and, presumably, also for electronic reasons. The resulting supramolecular structure is determined by the set of interactions that becomes dominant under given experimental conditions: in dilute solutions and with low-coordinating anions, the axle takes two wheels and one anion; in concentrated solutions and in dilute solutions with high-coordinating anions, the axle takes one wheel and two anions.

6.5   Conclusion

We have studied a new pseudorotaxane based on a tris(phenylureido)calix[6]arene wheel and a DAP-based cationic axle in apolar solvents. NMR and UV/Vis spectroscopic experiments evidenced the formation of an inclusion complex, in the shape of a pseudorotaxane. The wheel component is a heteroditopic receptor, which is able to complex both the cationic portion of the axle and its counterions. The axle shows characteristic spectroscopic properties that enabled us to investigate the interactions at the basis of the formation of the adducts by means of UV/Vis spectroscopic experiments. This study revealed how the nature of the counterions can affect not only the stability of an adduct, but also its stoichiometry.

References

 1. Balzani V, Venturi M, Credi A (2008) Molecular devices and machines—concepts and perspectives for the nano world, 2nd edn. Wiley-VCH, Weinheim
 2. Sauvage J-P, Dietrich-Buchecker C (eds) (1999) Catenanes, rotaxanes and knots. Wiley-VCH, Weinheim
 3. Hubin TJ, Busch DH (2000) Coord Chem Rev 200–202:5–52
 4. Mahan E, Gibson HW (2002) In: Semlyen JA (ed) Cyclic polymers, 2nd ed. Kluwer Publishers, Dordrecht, pp 415–560
 5. Panova IG, Topchieva IN (2001) Russ Chem Rev 70:23–44
 6. Takata T, Kihara N, Furusho Y (2004) Adv Polym Sci 171:1–75
 7. Huang F, Gibson HW (2005) Prog Polym Sci 30:982–1018
 8. Wenz G, Han B-H, Mueller A (2006) Chem Rev 106:782–817
 9. Faiz JA, Heitz V, Sauvage J-P (2009) Chem Soc Rev 38:422–442
10. Crowley JD, Goldup SM, Lee A-L, Leigh DA, McBurney RT (2009) Chem Soc Rev 38:1530–1541
11. Stoddart JF (2009) Chem Soc Rev 38:1802–1820
12. Harada A, Hashidzume A, Yamaguchi H, Takashima Y (2009) Chem Rev 109:5974–6023
13. Niu Z, Gibson HW (2009) Chem Rev 109:6024–6046
14. Reichardt C (1988) Solvents and solvent effects in organic chemistry. VCH, Weinheim
15. Isaacs N (1987) Physical organic chemistry. Longman, Essex, pp 49–55
16. Jones JW, Gibson HW (2003) J Am Chem Soc 125:7001–7004
17. Huang F, Jones JW, Slebodnick C, Gibson HW (2003) J Am Chem Soc 125:14458–14464
18. Gasa TB, Spruell JM, Dichtel WR, Sørensen TJ, Philp D, Stoddart JF, Kuzmic P (2009) Chem Eur J 15:106–116
19. Zhu K, Zhang M, Wang F, Li N, Li S, Huang F (2008) New J Chem 32:1827–1830
20. Clemente-León M, Pasquini C, Hebbe-Viton V, Lacour J, Dalla Cort A, Credi A (2006) Eur J Org Chem 105–112

21. Beer PD, Gale PA (2001) Angew Chem 113:502–532
22. Beer PD, Gale PA (2001) Angew Chem Int Ed 40:486–516
23. Gale PA (2003) Coord Chem Rev 240:191–221
24. Arduini A, Brindani E, Giorgi G, Pochini A, Secchi A (2002) J Org Chem 67:6188–6194
25. Hunter CA, Low CMR, Rotger C, Vinter JG, Zonta C (2003) Chem Commun 834–835
26. Casnati A, Massera C, Pelizzi N, Stibor I, Pinkassik E, Ugozzoli F, Ungaro R (2002) Tetrahedron Lett 43:7311–7314
27. Böhmer V, Dalla Cort A, Mandolini L (2001) J Org Chem 66:1900–1902
28. Hamon M, Ménand M, Le Gac S, Luhmer M, Dalla V, Jabin I (2008) J Org Chem 73: 7067–7071
29. Pescatori L, Arduini A, Pochini A, Secchi A, Massera C, Ugozzoli F (2009) CrystEngComm 11:239–241
30. Pescatori L, Arduini A, Pochini A, Secchi A, Massera C, Ugozzoli F (2009) Org Biomol Chem 7:3698–3708
31. Zhu K, Li S, Wang F, Huang F (2009) J Org Chem 74:1322–1328
32. Arduini A, Calzavacca F, Pochini A, Secchi A (2003) Chem Eur J 9:793–799
33. Arduini A, Bussolati R, Credi A, Faimani G, Garaudée S, Pochini A, Secchi A, Semeraro M, Silvi S, Venturi M (2009) Chem Eur J 15:3230–3242
34. Arduini A, Ciesa F, Fragassi M, Pochini A, Secchi A (2005) Angew Chem 117:282–285
35. Arduini A, Ciesa F, Fragassi M, Pochini A, Secchi A (2005) Angew Chem Int Ed 44:278–281
36. Sindelar V, Cejas MA, Raymo FM, Kaifer AE (2005) New J Chem 29:280–282
37. Sindelar V, Cejas MA, Raymo FM, Chen W, Parker SA, Kaifer AE (2005) Chem Eur J 11:7054–7059
38. Flood AH, Peters AJ, Vignon SA, Steuerman DW, Tseng H-R, Kang S, Heath JR, Stoddart JF (2004) Chem Eur J 10:6558–6564
39. Zhang X, Zhai C, Li N, Liu M, Li S, Zhu K, Zhang J, Huang F, Gibson HW (2007) Tetrahedron Lett 48:7537–7541
40. Doddi G, Ercolani G, Mencarelli P, Papa G (2007) J Org Chem 72:1503–1506
41. Ballardini R, Credi A, Gandolfi MT, Giansante C, Marconi G, Silvi S, Venturi M (2007) Inorg Chim Acta 360:1072–1082
42. Johnson CS Jr (1999) Prog Nucl Magn Reson Spectrosc 34:203–256 and references therein
43. Cohen Y, Avram L, Frish L (2005) Angew Chem 117:524–560
44. Cohen Y, Avram L, Frish L (2005) Angew Chem Int Ed 44:520–554
45. Avram L, Cohen Y (2002) J Org Chem 67:2639–2644
46. Binstead RA (1996) SPECFIT. Spectrum Software Associates, Chapel Hill
47. Job P (1928) Ann Chim 9:113–203
48. Gil VMS, Oliveira NC (1990) J Chem Educ 67:473–478
49. Schneider H-J, Yatsimirsky A (2000) Principles and methods in supramolecular chemistry. Wiley, New York
50. Credi A, Dumas S, Silvi S, Venturi M, Arduini A, Pochini A, Secchi A (2004) J Org Chem 69:5881–5887
51. Bianchi A, Bowman-James K, García-España E (eds) (1997) Supramolecular chemistry of anions. Wiley-VCH, Weinheim
52. Arduini A, Bussolati R, Credi A, Pochini A, Secchi A, Silvi S, Venturi M (2008) Tetrahedron 64:8279–8286

Chapter 7
A Simple Molecular Machine Operated by Photoinduced Proton Transfer

7.1 Introduction

Molecular machines—(supra)molecular systems in which large-amplitude motions of some components can be controlled by appropriate stimuli can be operated by means of chemical, electrochemical, or photochemical processes [1–3]. However, the use of light stimulation [4–6] has several advantages. For example, photons can be used to supply energy to the system and to gather information about its state without physically touching it. Light excitation can be carried out by a variety of sources (including the sun), with the possibility of a fine resolution in space and time.

Many artificial molecular machines reported so far are powered by exergonic chemical reactions, most typically acid–base reactions (for recent examples of pH-driven molecular machines, see: [7–12]). A modular construction of light-driven molecular machines, usually pursued by integrating photochemical functions with mechanically switchable systems [13–15], is in general more difficult to carry out. It would therefore be useful to identify viable strategies for using light to operate "stand alone" chemically driven molecular machines [16, 17]. Here we show that the acid–base controlled threading–dethreading of a pseudorotaxane in solution

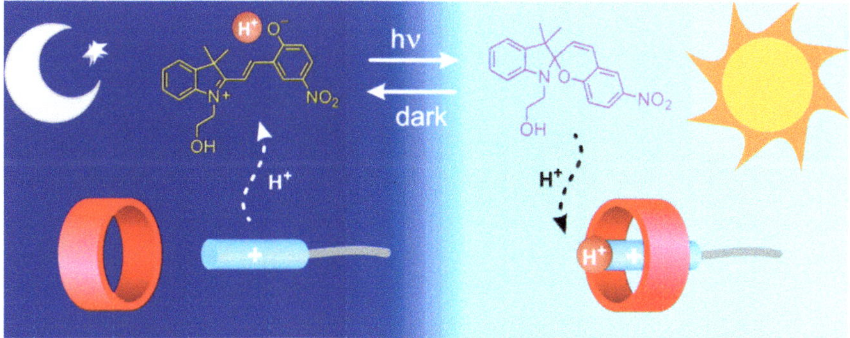

M. Baroncini, *Design, Synthesis and Characterization of new Supramolecular Architectures*, Springer Theses, DOI: 10.1007/978-3-642-19285-2_7,
© Springer-Verlag Berlin Heidelberg 2011

can be operated by photoinduced intermolecular proton transfer [18–20] with a molecular switch. Pseudorotaxanes whose molecular components can be threaded and dethreaded in response to external signals may be regarded as very simple prototypes of chemical machinery ([21], for interesting pseudorotaxane-based molecular devices, see: [22]) and are important for the development of less trivial unimolecular machines based on rotaxanes (recent example: [23]), catenanes (recent example: [24]), and related interlocked compounds [25–27].

7.2 Results and Discussion

The calix[6]arene wheel **C** (Scheme 7.1) forms fairly stable pseudorotaxane complexes with 4,4′-bipyridinium compounds in apolar solvents [28]. Therefore, we envisaged that compound $\mathbf{AH^{2+}}$, obtained by protonation of the pyridine

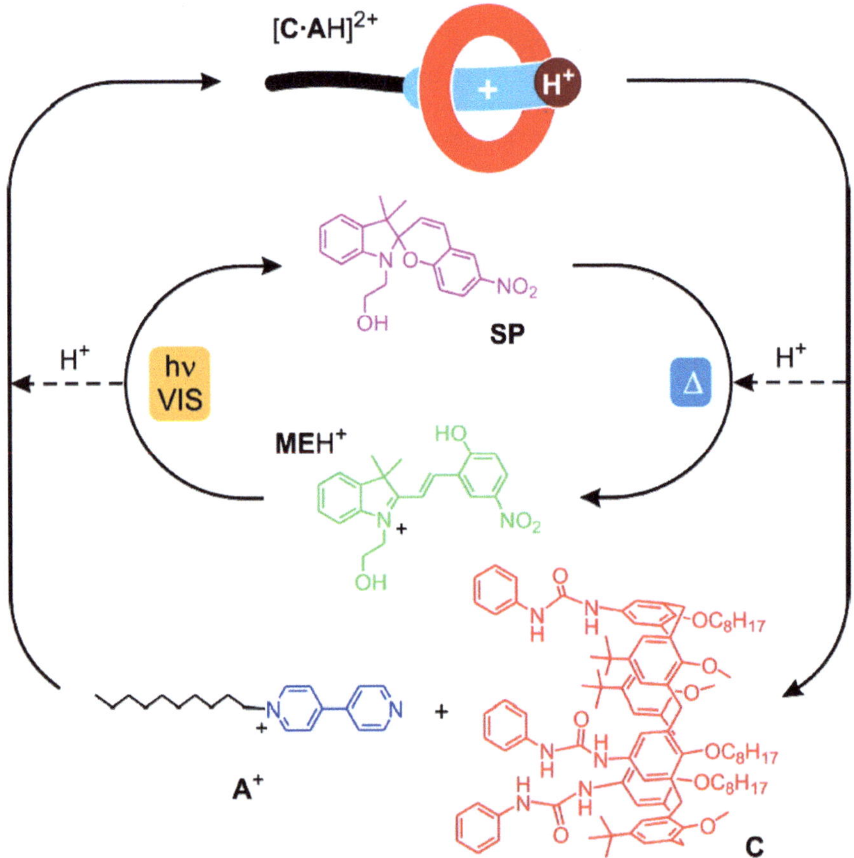

Scheme 7.1 Control of threading–dethreading processes in pseudorotaxane [C·AH]²⁺ by means of light-induced proton exchange with a spiropyran–merocyanine photochromic system

nitrogen of the 4,4$'$-pyridylpyridinium axle-like $\mathbf{A}^+$ (as PF_6^- salt; Scheme 7.1), could thread into the cavity of $\mathbf{C}$ as well. In fact, spectrophotometric titrations and voltammetric experiments show that a very stable $[K = (6 \pm 2) \times 10^6 \text{ M}^{-1}]$ pseudorotaxane complex is formed between $\mathbf{C}$ and $\mathbf{AH}^{2+}$ in CH_2Cl_2 (see Supporting Information). For the present discussion, it is important to note that the pseudorotaxane $[\mathbf{C \cdot AH}]^{2+}$ exhibits a broad and weak charge-transfer (CT) absorption band in the visible region ($\lambda_{max} = 478$ nm, $\varepsilon = 500 \text{ M}^{-1} \text{ cm}^{-1}$), where none of the isolated molecular components exhibit absorption features. Deprotonation of $\mathbf{AH}^{2+}$ with a base (e.g., tributylamine) in CH_2Cl_2 leads to dethreading of the pseudorotaxane.

In order to trigger the self-assembly and disassembly of this pseudorotaxane by light, a suitable species whose acid–base properties can be photocontrolled should be identified. Specifically, such a compound must occur in two forms, interconvertible into one another by light irradiation, exhibiting smaller and larger acid strength than that of $\mathbf{AH}^{2+}$, respectively. A species that fulfils these requirements is the spiropyran photochrome $\mathbf{SP}$ (Scheme 7.1; [29], recent examples of photoswitchable systems based on spiropyrans: [30–33]). In the presence of an acid, the colorless $\mathbf{SP}$ is converted into the yellow protonated merocyanine form $\mathbf{MEH}^+$ [34]. Upon irradiation with visible light, $\mathbf{MEH}^+$ releases a proton, isomerizing back to $\mathbf{SP}$ [35, 36]. The coupled operation expected for the pseudorotaxane and spiropyran switches is represented in Scheme 7.1.

The absorption spectrum of a 1:1 mixture of the protonated axle $\mathbf{AH}^{2+}$ and the calixarene wheel $\mathbf{C}$ shows the typical [28] CT absorption band of the $[\mathbf{C \cdot AH}]^{2+}$ complex (Fig. 7.1, black curve). In our conditions (1×10^{-4} M), more than 95% of the molecular components are associated together in the pseudorotaxane superstructure. The CT band of $[\mathbf{C \cdot AH}]^{2+}$ does not change soon after the addition of 1 equiv of $\mathbf{SP}$ (blue curve). By keeping the solution in the dark, the absorption band typical [35, 36] of $\mathbf{MEH}^+$ at 417 nm gradually appears. Although the $\mathbf{MEH}^+$ band partially overlaps with the much weaker CT band of the complex, it can be noticed that the absorbance at $\lambda > 520$ nm decreases concomitantly with the formation of $\mathbf{MEH}^+$ (Fig. 7.1, red curve). This change cannot be ascribed to the $\mathbf{SP}$-$\mathbf{MEH}^+$ transformation and has to be assigned to the dethreading of the $[\mathbf{C \cdot AH}]^{2+}$ pseudorotaxane. When the equilibration is completed (after 7 days), it can be estimated on the basis of the $\mathbf{MEH}^+$ absorption band at 417 nm that the $\mathbf{SP}$:$\mathbf{MEH}^+$ ratio is about 60:40; however, the decrease of the CT absorption would correspond to disassembling of 15% of the $[\mathbf{C \cdot AH}]^{2+}$ species, instead of the 40% expected from the amount of $\mathbf{MEH}^+$ formed. Most likely, the decrease in the CT band is partially offset by the absorbance increase originating from the formation of a small amount of the nonprotonated merocyanine $\mathbf{ME}$ (which absorbs strongly in the 500–600 nm region) [34–36] in equilibrium with $\mathbf{MEH}^+$. By irradiating the solution with visible light, the $\mathbf{MEH}^+$ band disappears completely, and the initial CT band is fully restored (Fig. 7.1, green curve). The inset of Fig. 7.1 shows the absorbance changes observed at 417 nm ($\mathbf{MEH}^+$ band) and 550 nm (pseudorotaxane CT band) by repeated thermal equilibration–light irradiation cycles, indicating that the switching process is fully reversible.

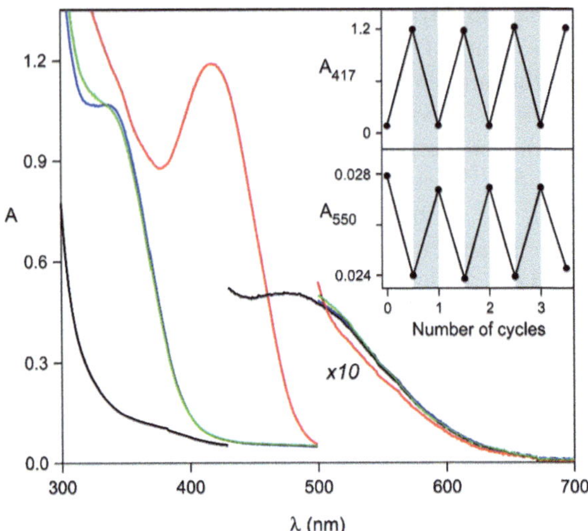

Fig. 7.1 Absorption spectra in CH_2Cl_2 at room temperature of (i) the pseudorotaxane $[C \cdot AH]^{2+}$, black curve; (ii) $[C \cdot AH]^{2+}$ and **SP**, immediately after the addition of the latter, blue curve; (iii) solution (ii) after 7 days of rest in the dark, red curve; (iv) solution (iii) after 10 min of irradiation at $\lambda > 450$ nm, green curve. *Inset*: absorbance changes at 417 nm (a) and 550 nm (b) of a CH_2Cl_2 solution containing $[C \cdot AH]^{2+}$ and **SP** observed upon several cycles of thermal equilibration (7 days in the dark at room temperature, white areas) and visible light irradiation (10 min with a 150 W tungsten–halogen lamp, $\lambda > 450$ nm, gray areas). The concentration of all the components was 1.0×10^{-4} M

The observed changes can be interpreted in terms of the reactions shown in Scheme 7.1. Starting from a mixture of the complex $[C \cdot AH]^{2+}$ and **SP**, a thermal proton transfer occurs from AH^{2+} to the photochrome, yielding MEH^+ and the deprotonated axle A^+ which is not appreciably complexed by **C** in our conditions. Hence, a slow dark reaction that leads to dethreading of the pseudorotaxane takes place. Subsequent light irradiation of the MEH^+ species in the visible region causes an opposite proton transfer converting A^+ into AH^{2+}, which rethreads into the calixarene wheel. Since the reset of the system occurs thermally, its operation under continuous light irradiation can give rise to autonomous behavior; in other words, the dethreading–rethreading process is repeated without intervention of an operator as long as the energy source (i.e., light) is available. In practice, because of the large difference in the time scale of the dark and light parts of the cycle, the photostationary state is strongly displaced toward the $SP\text{-}[C \cdot AH]^{2+}$ state, unless irradiation is carried out with very low intensity and/or the temperature is increased. Nevertheless, this behavior could be employed to implement a memory effect in the system [18–20].

7.3 Conclusion

In conclusion, we have shown that the threading–dethreading of a pH-switchable calix[6]arene bipyridinium pseudorotaxane in solution can be controlled by visible light irradiation using a merocyanine compound as a photoacid. Such a coupling between the two molecular switches that communicate with one another by intermolecular proton transfer provides a general principle for the operation of photoinactive acid–base controllable molecular machines with light. Investigations aimed at elucidating the switching mechanism in more detail, improving the performances of the system, and applying this strategy to more complex nanodevices are underway.

References

1. Balzani V, Credi A, Venturi M (2003) Molecular devices and machines. Wiley-VCH, Weinheim
2. Browne WR, Feringa BL (2006) Nat Nanotechnol 1:25–35
3. Kay ER, Leigh DA, Zerbetto F (2007) Angew Chem Int Ed 46:72–191
4. Credi A (2006) Aust J Chem 59:157–169
5. Tian H, Wang QC (2006) Chem Soc Rev 35:361–364
6. Saha S, Stoddart JF (2007) Chem Soc Rev 36:77–92
7. Cheng K-W, Lai C-C, Chiang P-T, Chiu S-H (2006) Chem Commun 2854–2856
8. Ooya T, Inoue D, Choi HS, Kobayashi Y, Loethen S, Thompson DH, Ko YH, Kim K, Yui N (2006) Org Lett 8:3159–3162
9. Nguyen TD, Leung KCF, Liong M, Pentecost CD, Stoddart JF, Zink JI (2006) Org Lett 8:3363–3366
10. Leigh DA, Thomson AR (2006) Org Lett 8:5377–5379
11. Koo C-K, Lam B, Leung S-K, Lam MH-W, Wong W-YJ (2006) Am Chem Soc 128:16434–16435
12. Sindelar V, Silvi S, Parker SE, Sobransingh D, Kaifer AE (2007) Adv Funct Mater 17:694–701
13. Balzani V, Clemente-León M, Credi A, Ferrer B, Venturi M, Flood AH, Stoddart JF (2006) Proc Natl Acad Sci USA 103:1178–1183
14. Muraoka T, Kinbara K, Aida T (2006) Nature 440:512–515
15. Saha S, Flood AH, Stoddart JF, Impellizzeri S, Silvi S, Venturi M, Credi AJ (2007) Am Chem Soc 129:12159–12171
16. Liu H, Xu Y, Li F, Yang Y, Wang W, Song Y, Liu D (2007) Angew Chem Int Ed 46:2515–2517
17. Wu H, Zhang D, Su L, Ohkubo K, Zhang C, Yin S, Mao L, Shuai Z, Fukuzumi S, Zhu DJ (2007) Am Chem Soc 129:6839–6846
18. Raymo FM, Giordani S (2001) Org Lett 3:3475–3478
19. Raymo FM, Alvarado RJ, Giordani S, Cejas MAJ (2003) Am Chem Soc 125:2361–2364
20. Giordani S, Cejas MA, Raymo FM (2004) Tetrahedron 60:10973–10981
21. Balzani V, Credi A, Venturi M (2002) Proc Natl Acad Sci USA 99:4814–4817
22. Saha S, Leung KC-F, Nguyen TD, Stoddart JF, Zink JI (2007) Adv Funct Mater 17:685–693
23. Serreli V, Lee C-F, Kay ER, Leigh DA (2007) Nature 445:523–527
24. Ikeda T, Saha S, Aprahamian I, Leung KC-F, Williams A, Deng WQ, Flood AH, Goddard WA, Stoddart JF (2007) Chem Asian J 2:76–93
25. Jimenez-Molero MC, Dietrich-Buchecker C, Sauvage J-P (2002) Chem Eur J 8:1456–1466
26. Liu Y, Flood AH, Bonvallett PA, Vignon SA, Northrop BH, Tseng H-R, Jeppesen JO, Huang TJ, Brough B, Baller M, Magonov S, Solares SD, Goddard WA, Ho C-M, Stoddart JFJ (2005) Am Chem Soc 127:9745–9759
27. Badjic JD, Ronconi CM, Stoddart JF, Balzani V, Silvi S, Credi AJ (2006) Am Chem Soc 128:1489–1499

28. Credi A, Dumas S, Silvi S, Venturi M, Arduini A, Pochini A, Secchi AJ (2004) Org Chem 69:5881–5887
29. Dürr H, Bouas-Laurent H (eds) (2003) Photochromism: molecules and systems. Elsevier, Amsterdam
30. Andreasson J, Straight SD, Kodis G, Park CD, Hambourger M, Gervaldo M, Albinsson B, Moore TA, Moore AL, Gust D (2006) J Am Chem Soc 128:16259–16265
31. Wen GY, Yan J, Zhou YC, Zhang DQ, Mao LQ, Zhu DB (2006) Chem Commun 3016–3018
32. Zhu LY, Wu WW, Zhu MQ, Han JJ, Hurst JK, Li ADQJ (2007) Am Chem Soc 129:3524–3525
33. Niazov T, Shlyahovsky B, Willner IJ (2007) Am Chem Soc 129:6374–6375
34. Wojtyk JTC, Wasey A, Xiao N-N, Kazmaier PM, Hoz S, Yu C, Lemieux RP, Buncel EJ (2007) Phys Chem A 111:2511–2516
35. Raymo FM, Giordani SJ (2001) Am Chem Soc 123:4651–4652
36. Raymo FM, Giordani S, White AJP, Williams DJJ (2003) Org Chem 68:4158–4169

Chapter 8
Reversible Photoswitching of Rotaxane Character

8.1 Introduction

Pseudorotaxanes complexes can be rendered kinetically inert, i.e. transformed into rotaxanes, by attaching bulky groups at the extremities of the axle to prevent dethreading [1]. Therefore, the pseudorotaxane or rotaxane behavior of a given axle-macrocycle pair depends on the threading–dethreading rate constants, which in turn are determined by the intrinsic kinetic parameters of such processes— particularly, the energy barriers—and the experimental conditions [2, 3].

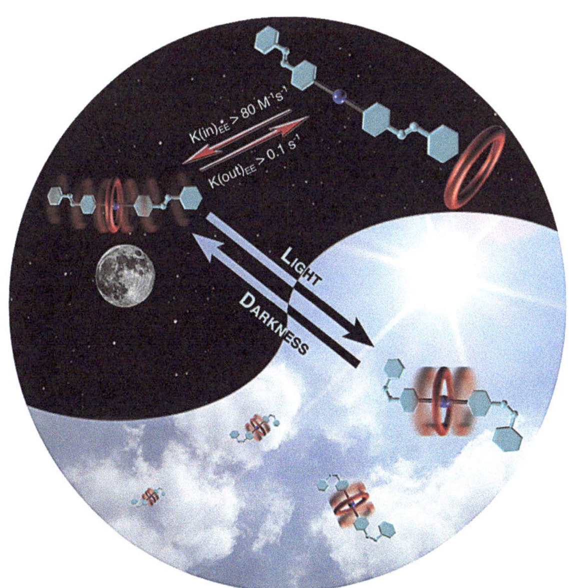

M. Baroncini, *Design, Synthesis and Characterization of new Supramolecular Architectures*, Springer Theses, DOI: 10.1007/978-3-642-19285-2_8,
© Springer-Verlag Berlin Heidelberg 2011

Rotaxanes and related species are primarily interesting for the construction of molecular machines [4–6]. The operation of most rotaxane-type machines is based on classical switching processes between thermodynamically stable states. It is clear, however, that functional molecular motors will only be realized if the reaction rates between the different states can also be controlled [7–9], thus enabling the implementation of ratchet-type mechanisms [10–12]. In this context, the ability of adjusting the threading–dethreading kinetics [13, 14] by modulating the corresponding energy barriers through external stimulation is an important achievement.

In this chapter is described a self-assembling system that can be switched between thermodynamically stable (pseudorotaxane) and kinetically inert (rotaxane) forms by light irradiation. The described system is advantageous because (1) it can be controlled by light, (2) it exhibits a clear-cut behavior, (3) it is fully reversible, and (4) it is structurally simple and easy to make [15–17]. It is also described how a change in the stability of the complex affects its kinetic behavior.

8.2 Results and Discussion

The system is composed by the bis-azobenzylamine axle EE-$1H^+$ and the dibenzo[24]crown-8 ether ring **2**. The axle was synthesized in good yields by Mill's coupling of bis(4-aminobenzyl)amine and 4-nitrosotoluene in acetic acid, followed by anion exchange with NH_4PF_6. The EE-$1H \cdot PF_6$ salt was fully characterized by ^1H and ^{13}C NMR, DQF-COSY, ESI–MS, and UV–visible absorption spectroscopies. The ^1H NMR spectrum of a 1:1 mixture of EE-$1H^+$ and **2** in CD_3CN at room temperature shows sharp and well resolved signals associated with the 1:1 complex $[EE\text{-}1H \subset 2]^+$ (Scheme 8.1) in equilibrium with the free axle and ring components. The NMR data are fully consistent with a pseudorotaxane structure in which the crown ether surrounds the ammonium center of the axle [18, 19].

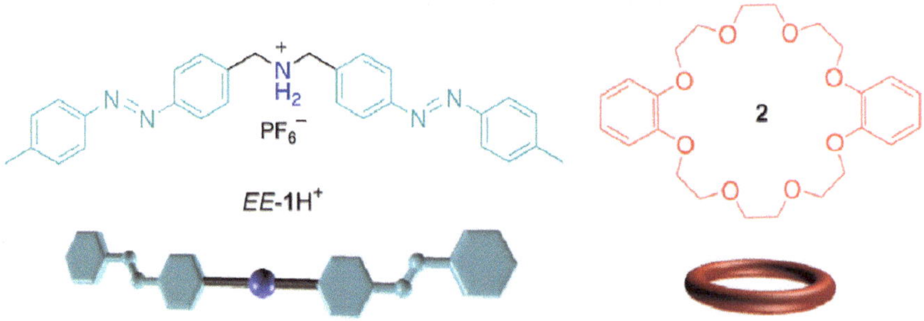

EE-$1H^+$

2

The stability constant of $[EE\text{-}1H \subset 2]^+$ (K_{EE}) was obtained in different solvents at 298 K by single-point determination from ^1H NMR spectra. In acetonitrile, which is the solvent of choice for all the successive experiments, $K_{EE} = 820\ M^{-1}$. The equilibrium between the complex and its free components was established

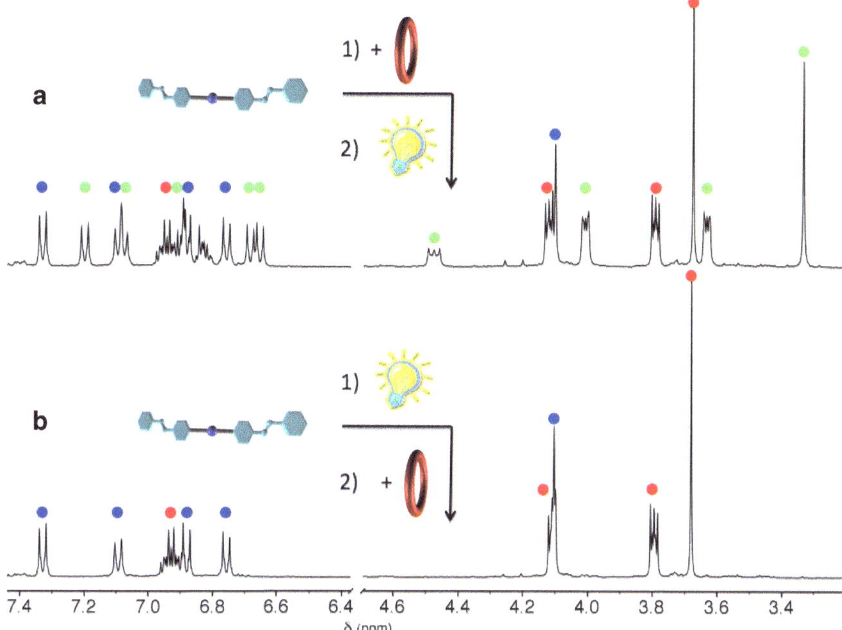

Fig. 8.1 ^{1}H NMR spectra (400 MHz, CD$_3$CN, 298 K, 3 mM) of *EE*-**1H**$^+$ after: **a** addition of 1 equivalent of **2** followed by exhaustive irradiation at 365 nm; **b** same inputs in the reverse order. *Blue*, *red* and *green* marks denote resonances of **1H**$^+$, **2** and their complex, respectively

within the time necessary for mixing and recording the NMR spectrum. Hence, we can estimate a lower-limiting value for the threading rate constant $k(\text{in})_{EE}$ of 5 M^{-1} s^{-1} and, accordingly, $k(\text{out})_{EE}$ 6 × 10^{-3} s^{-1}.

Photoirradiation of *EE*-**1H**$^+$ at 365 nm afforded almost quantitative (> 95%) isomerization to the *ZZ* isomer, as shown by ^{1}H NMR and UV–vis spectra. The values determined for the *EZ* photo-isomerization quantum yield (0.095) and rate constant for *ZE* dark isomerization (1.4 × 10^{-6} s^{-1}) are typical of the azobenzene unit and are virtually unaffected by the presence of ring **2**. Thus, both the end groups of **1H**$^+$—either alone or surrounded by **2**—can be effectively switched between the *E* and *Z* isomers. Cycling between *EE* and *ZZ* states could be performed with no sign of degradation. The ^{1}H NMR spectra also showed that *ZZ*-**1H**$^+$ remains complexed by **2** (Fig. 8.1a).

The inspection of molecular models suggested that the *Z*-azobenzene unit could exert a stronger steric hindrance compared with *E*-azobenzene for the threading of **2** on **1H**$^+$. Indeed, no sign of complexation was observed by ^{1}H NMR soon after 1:1 mixing of **2** and *ZZ*-**1H**$^+$ in CD$_3$CN (Fig. 8.1b). However, very slow changes, consistent with the formation of the [*ZZ*-**1H** ⊂ **2**]$^+$ complex, took place (Fig. 8.2). A kinetic analysis of the NMR spectra afforded a rate constant $k(\text{in})_{ZZ} = 2.9 \times 10^{-3}$ M^{-1} s^{-1} while the stability constant, determined at equilibrium, resulted to be $K_{ZZ} = 400$ M^{-1}. From these data we calculated a dethreading rate

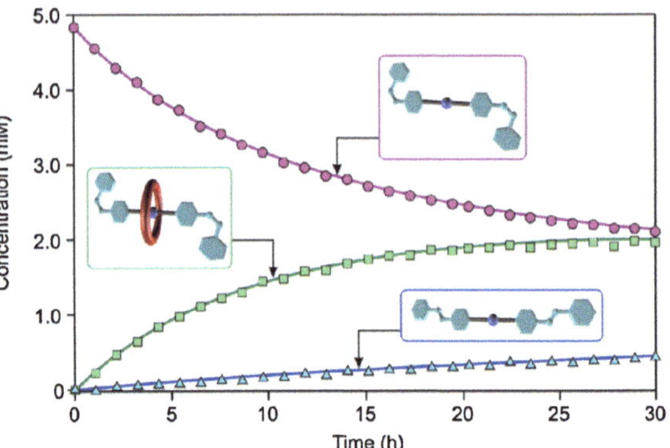

Fig. 8.2 Concentration time profiles, obtained from NMR data, of free and complexed $ZZ\mathbf{1}H^+$ upon mixing 4.8 mM $ZZ\mathbf{1}H^+$ and 5.4 mM **2** (CD$_3$CN, 298 K). The thermal regeneration of $EE\mathbf{1}H^+$ is also shown. The *solid lines* represent the data fitting according to the model of Scheme 8.1

constant for the ZZ isomer, $k(\text{out})_{ZZ} = 7.2 \times 10^{-6}$ s^{-1}. Therefore, the disassembly of the $[ZZ\text{-}\mathbf{1}H \subset \mathbf{2}]^+$ complex is three orders of magnitude slower than that of $[EE\text{-}\mathbf{1}H \subset \mathbf{2}]^+$ and takes place on the same timescale of the ZE thermal isomerization. In other words, light irradiation transforms the $[EE\text{-}\mathbf{1}H \subset \mathbf{2}]^+$ pseudorotaxane into a rotaxane species which remains kinetically interlocked for days (Scheme 8.1).

Now the question is: can Z-azobenzene units prevent dethreading if the intercomponent interactions between the ring and the axle are removed? Our system is

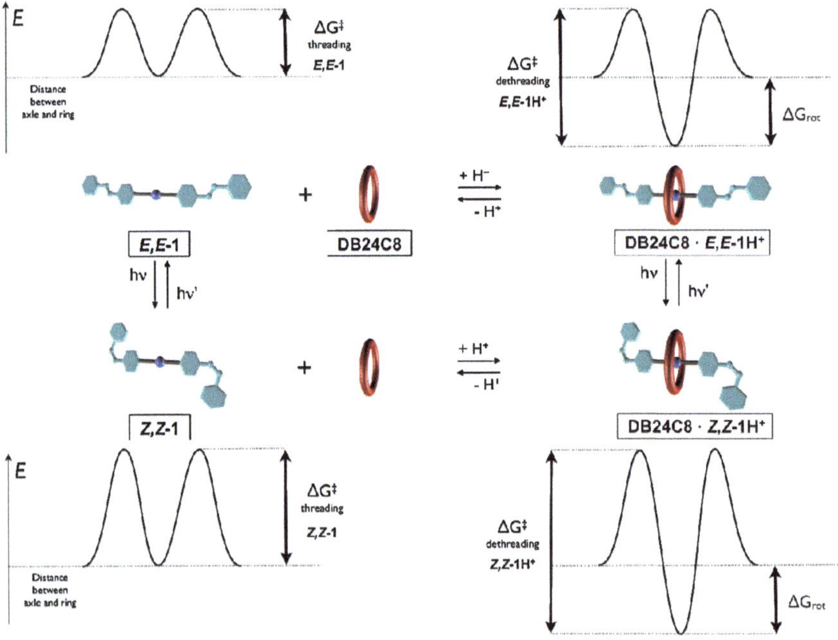

Scheme 8.1 Representation of the chemical equilibria and photochemical reactions involving components $\mathbf{1}H^+$ and **2** and simplified potential energy curves (free energy versus ring-axle distance) for the threading–dethreading of the ring and the axle in its EE (*left*) and ZZ (*right*) isomeric forms

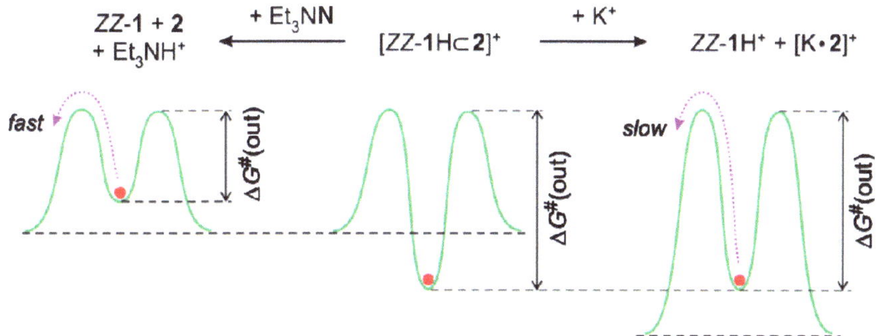

Scheme 8.2 Simplified potential energy curves showing the dethreading of the ring from the *ZZ* axle upon deprotonation of the ammonium center of the axle with triethylamine (*left*) or sequestration of the ring with K^+ as a competitive guest (*right*)

particularly appropriate for investigating this aspect because such interactions can be virtually cancelled by deprotonating the ammonium center of the axle with a base, i.e. using a stimulus orthogonal to that employed for switching the azobenzene end units. Interestingly, deprotonation of the ammonium center of $[ZZ\text{-}1H \subset 2]^+$ with triethylamine resulted in a quantitative and fast dethreading, as in the case of $[EE\text{-}1H \subset 2]^+$. This observation indicates that the dethreading barrier is dramatically lowered by the thermodynamic destabilization of the threaded state (Scheme 8.2, left). Such an effect is a consequence of the proximity of the ammonium center and azobenzene pseudo-stoppers.

In the attempt of dethreading the $[ZZ\text{-}1H \subset 2]^+$ species without weakening the intercomponent interactions, we added K^+ ions as a competitive guest for crown ether **2**. Indeed, the addition of 2.9 equivalents of KPF_6 to the $[EE\text{-}1H \subset 2]^+$ pseudorotaxane caused its complete and fast disassembly. Conversely, $[ZZ\text{-}1H \subset 2]^+$ disassembled very slowly upon addition of 2.9 equivalents of KPF_6. The time-dependent NMR spectral changes show that dethreading occurs on a time scale comparable to that observed in the absence of K^+ (vide supra). This behavior can be interpreted on the basis of the energy curves shown in the right part of Scheme 8.2.

8.3 Conclusion

In summary, we have shown that azobenzene units can be effectively employed as end groups in a molecular axle to implement photocontrol on the threading–dethreading rate with a crown ether ring. The control of the kinetics of threading–dethreading in rotaxane system is an important achievement to elaborate more complex function such as the development of non-symmetric axle-ring systems in which it is possible to control the threading direction by light irradiation.

References

1. Sauvage J-P, Dietrich-Buchecker C (eds) (1999) Molecular catenanes, rotaxanes and knots: a journey through the world of molecular topology. Wiley-VCH, Weinheim
2. Ashton PR, Baxter I, Fyfe MCT, Raymo FM, Spencer N, Stoddart JF, White AJP, Williams DJJ (1998) Am Chem Soc 120:2297–2307
3. Bolla MA, Tiburcio J, Loeb SJ (2008) Tetrahedron 64:8423–8427
4. Browne WR, Feringa BL (2006) Nat Nanotech 1:25–35
5. Kay ER, Leigh DA, Zerbetto F (2007) Angew Chem Int Ed 46:72–191
6. Balzani V, Credi A, Venturi M (2008) Molecular devices and machines—concepts and perspectives for the nano world. Wiley-VCH, Weinheim
7. Chen N-C, Lai C-C, Liu Y-H, Peng S-M, Chiu S-H (2008) Chem Eur J 14:2904–2908
8. Coskun A, Friedman DC, Li H, Patel K, Khatib HA, Stoddart JFJ (2009) Am Chem Soc 131:2493–2495
9. Share AI, Parimal K, Flood AHJ (2010) Am Chem Soc 132:1665–1675
10. Chatterjee MN, Kay ER, Leigh DA (2006) J Am Chem Soc 128:4058–4073
11. Serreli V, Lee C-F, Kay ER, Leigh DA (2007) Nature 445:523–527
12. Astumian RD (2007) Phys Chem Chem Phys 9:5067–5083
13. Oshikiri T, Takashima Y, Yamaguchi H, Harada A (2007) Chem Eur J 13:7091–7098
14. Arduini A, Bussolati R, Credi A, Faimani G, Garaudee S, Pochini A, Secchi A, Semeraro M, Silvi S, Venturi M (2009) Chem Eur J 15:3230–3242
15. Jiang L, Okano J, Orita A, Otera J (2004) Angew Chem Int Ed 43:2121–2124
16. Hirose K, Shiba Y, Ishibashi K, Doi Y, Tobe Y (2008) Chem Eur J 14:981–986
17. Tokunaga Y, Akasaka K, Hashimoto N, Yamanaka S, Hisada K, Shimomura Y, Kakuchi SJ (2009) Org Chem 74:2374–2379
18. Cantrill SJ, Fulton DA, Heiss AM, Pease AR, Stoddart JF, White AJP, Williams DJ (2000) Chem Eur J 6:2274–2287
19. Jones JW, Gibson HWJ (2003) Am Chem Soc 125:7001–7004

Chapter 9
A Simple Unimolecular Multiplexer/Demultiplexer

9.1 Introduction

In living organisms information is processed, transferred, and stored using molecular or ionic substrates [1]. The design and construction of molecular systems capable of elaborating information [2, 3] could lead to the development of radically new computing paradigms and, in the long term, to the realization of useful devices. Leaving aside speculation related to the construction of a chemical computer, molecular logic systems could perform relatively simple computing tasks that cannot be accomplished with silicon-based devices, e.g. in nanoscale spaces [4] or in vivo [5]. Molecular switches convert input stimuli into output signals [6]. Hence, the principles of binary logic [7] can be applied to signal transduction operated by molecules under appropriate conditions (a large number of chemical switches exhibiting interesting properties from a binary logic viewpoint have been reported. In many cases the authors were either not aware of such behavior in their systems, or they published the work before the notion of molecular logic gates became popular. For the first explicit description of the analogy between molecular switches and logic gates, see: [8]). Many chemical systems that mimic the operation of semiconductor logic gates and circuits have been reported [9–28]. An important function in information technology is signal multiplexing/demultiplexing. A 2:1 multiplexer (mux) is a circuit with two data inputs, one address input, and one output. The mux selects the binary state from one of the data inputs and directs it to the output; the selected input depends on the

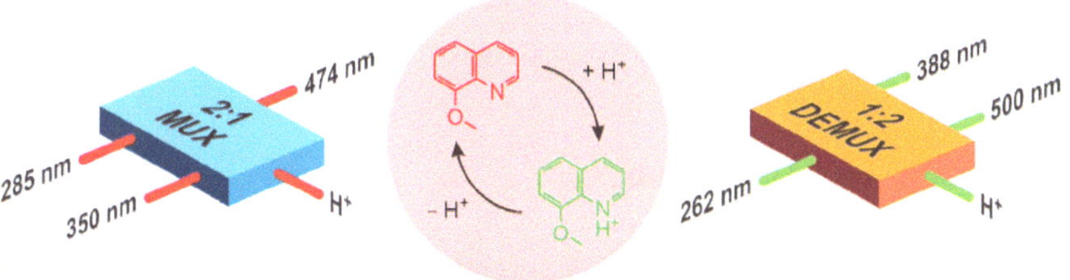

M. Baroncini, *Design, Synthesis and Characterization of new Supramolecular Architectures*, Springer Theses, DOI: 10.1007/978-3-642-19285-2_9,
© Springer-Verlag Berlin Heidelberg 2011

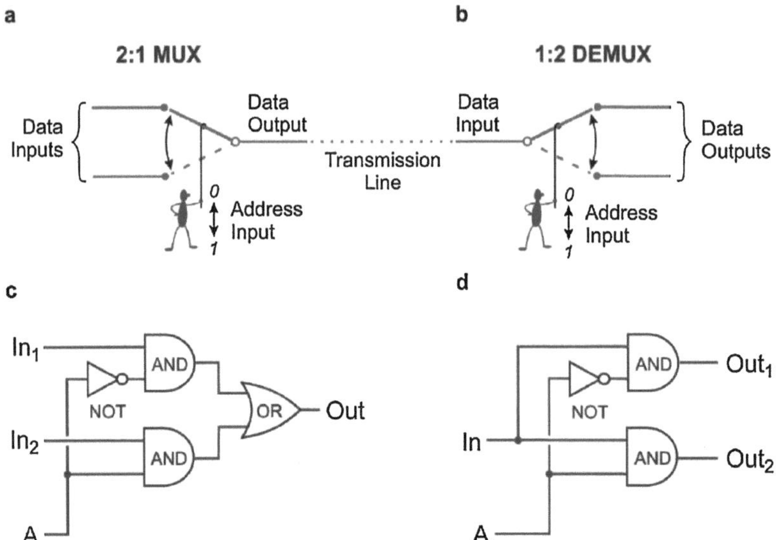

Scheme 9.1 Operation scheme (*top*) and combinational logic network (*bottom*) corresponding to a 2:1 multiplexer (**a** and **c**) and a 1:2 demultiplexer (**b** and **d**)

binary state of the address input (Scheme 9.1). Conversely, a 1:2 demultiplexer (demux) is a circuit that possesses one data input, one address input, and two outputs. The demux routes the data input to one of the output lines, and the selected output is determined by the binary state of the address input (Scheme 9.1b). Hence a multiplexer allows the encoding of multiple data streams into a single data line for transmission, and a demultiplexer can decode entangled data streams from a single signal (Scheme 9.1a, b). The combinational circuits corresponding to a 2:1 multiplexer and 1:2 demultiplexer are shown in Scheme 9.1. Molecules that can mimic the function of a 2:1 multiplexer [29, 30] or a 1:2 demultiplexer [31, 32], have recently been reported. However, these systems either rely on carefully designed multicomponent species [29, 33, 34], and coupling to an external optical device [31], or imply a dependence of the data input on the binary state of the address input [32]. Demultiplexers based on solid-state nanostructures have also been described [33, 34].

Here we show that the reversible acid/base switching of the absorption and photoluminescence properties of a fluorophore as simple as 8-methoxyquinoline in solution can form the basis for molecular 2:1 multiplexing and 1:2 demultiplexing with a clear-cut digital response.

9.2 Results and Discussion

8-Methoxyquinoline (8-MQ, Scheme 9.2) is related to the well-known fluorogenic compound 8-hydroxyquinoline (8-HQ) (for examples of molecular switches based

Scheme 9.2 Acid/base-controlled switching between 8-methoxyquinoline (8-MQ) and its protonated form (8-MQ-H$^+$)

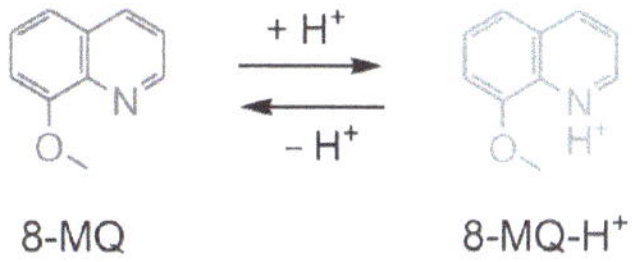

on 8-MQ derivatives, see: [35–37]). The absorption spectrum of 8-MQ in CH$_3$CN (Fig. 9.1a) shows a band with $\lambda_{max} = 301$ nm, assigned to a π–π^* transition [38–40]. In contrast with 8-HQ, 8-MQ is strongly fluorescent ($\lambda_{max} = 388$ nm, Fig. 9.1b) because the presence of the methoxy group prevents the intramolecular proton transfer process in the excited state responsible for the strong fluorescence quenching in 8-HQ [41]. The addition of one equivalent of triflic acid (CF$_3$SO$_3$H) to 8-MQ affords the protonated form 8-MQ-H$^+$ (Scheme 9.2), the absorption and fluorescence spectra of which are markedly different from those of 8-MQ. Specifically, the absorbance in the region of the 8-MQ absorption band (270–320 nm) decreases substantially, and a new absorption band ($\lambda_{max} = 358$ nm, Fig. 9.1c) is observed in a spectral region where 8-MQ does not absorb light. The fluorescence band of 8-MQ, which has a maximum at 388 nm, is replaced by a weaker emission band with $\lambda_{max} = 500$ nm (Fig. 9.1d). The unprotonated form can be regenerated on addition of one equivalent of tris-n-butylamine (nBu$_3$N). The acid/base-controlled switching between 8-MQ and 8-MQ-H$^+$ is fully reversible and can be repeated many times with the same solution without appreciable reduction in intensity in the absorption and fluorescence spectra. This particular spectroscopic behavior and the chemical reversibility of the acid/base switching can be used to obtain both 2:1 mux and 1:2 demux functions, using proton concentration as the address input (A), and excitation and emission optical signals as the data inputs and outputs, respectively.

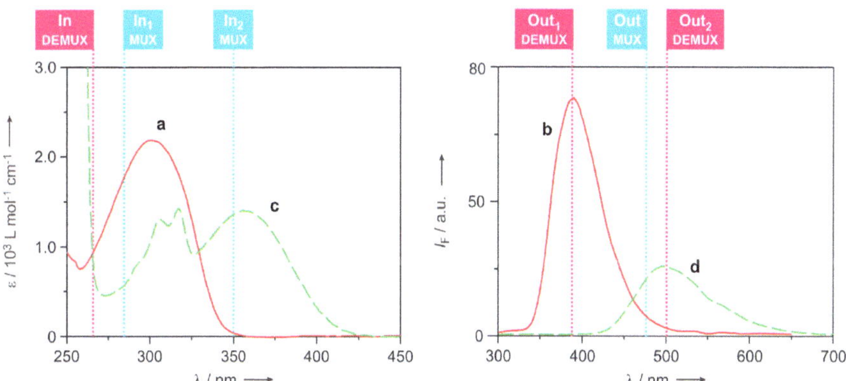

Fig. 9.1 Absorption (*top*) and fluorescence ($\lambda_{ex} = 262$ nm, *bottom*) spectra of 8-MQ (*a* and *b*) and 8-MQ-H$^+$ (*c* and *d*). The wavelengths of the inputs and outputs signals relevant for the mux/demux binary functions are indicated. Conditions: 1.5×10^{-5} mol L^{-1}, air-equilibrated MeCN, RT

Table 9.1 Truth table for the 2:1 multiplexer function

Data inputs		Address input	Data output
In_1	In_2	A	Out
I_{ex} (285 nm)[a]	I_{ex} (350 nm)[a]	$[H^+]/[8MQ]$	I_F (474 nm)[b]
0	0	0	0 (0)
1	0	0	1 (21)
0	1	0	0 (5.1)
1	1	0	1 (26)
0	0	1	0 (0)
1	0	1	0 (9.2)
0	1	1	1 (21)
1	1	1	1 (30)

[a] Binary state 1 corresponds to irradiation with the excitation lamp of the spectrofluorimeter; state 0 corresponds to no excitation (lamp off)

[b] The values in parentheses indicate the experimental fluorescence intensity values in arbitrary units; the corresponding binary states are determined by applying a threshold value of $I_F = 15$ arbitrary units

In the case of the 2:1 multiplexer the incident light intensities at 285 nm (In_1) and 350 nm (In_2) were used as the data inputs; these wavelengths were chosen in order to achieve selective excitation of 8-MQ and 8-MQ-H$^+$, respectively (Fig. 9.1a–c). The output (Out) was monitored using the fluorescence intensity at 474 nm, a wavelength at which both 8-MQ and 8-MQ-H$^+$ fluoresce (Fig. 9.1b–d). The fluorescence output levels measured for a solution containing 8-MQ in MeCN under the conditions corresponding to the eight combinations of the binary data and address inputs are listed in Table 9.1. A threshold value could be easily identified in order to assign binary output states so that the truth table corresponds to that of a 2:1 multiplexer. If the address input is 0 (no H$^+$ added, 8-MQ form), the output reports the state of data input In_1, whereas if the address input is 1 (1 equivalent of H$^+$ added, 8-MQ-H$^+$ form), the output mirrors the state of data input In_2.

The system can be easily reconfigured to behave as a 1:2 demultiplexer by changing the optical input and output channels. In this case, the data input (In) was the incident light intensity at 262 nm, which is an isosbestic point in the absorption spectra of 8-MQ and 8-MQ-H$^+$. The two output signals were represented by the fluorescence intensity at 388 nm (λ_{max} for 8-MQ, Out_1) and 500 nm (λ_{max} for 8-MQ-H$^+$, Out_2), respectively. Table 9.2 shows the fluorescence output levels measured for a solution of 8-MQ in MeCN under the conditions corresponding to the four combinations of the binary data and address inputs. On fixing an appropriate threshold for the fluorescence output, the truth table of the 1:2 demultiplexer is obtained. If the address input is 0 (8-MQ form), the binary state of the data input is transmitted to Out_1; conversely, if the address input is 1 (8-MQ-H$^+$ form), the binary data input is transmitted to Out_2. Although the construction of a practical device was not the overall purpose of this work, we investigated whether the acid/base switching of the 8-MQ fluorescence observed in solution could also be obtained in a solid-state environment.

Table 9.2 Truth table for the 1:2 demultiplexer function

Data input	Address input	Data outputs	
In	*A*	*Out*$_1$	*Out*$_2$
I_{ex} (262 nm)[a]	$[H^+]$/[8MQ]	I_F (388 nm)[b]	I_F (500 nm)[b]
0	0	0 (0)	0 (0)
1	0	1 (71)	0 (0.5)
0	1	0 (0)	0 (0)
1	1	0 (2.1)	1 (21)

[a] Binary state 1 corresponds to irradiation with the excitation lamp of the spectrofluorimeter; state 0 corresponds to no excitation (lamp off)

[b] The values in parentheses indicate the experimental fluorescence intensity values in arbitrary units; the corresponding binary states are determined by applying a threshold value of $I_F = 10$ arbitrary units

To this end, we embedded 8-MQ into polystyrene thin films and recorded the fluorescence spectra of the resulting materials upon exposure to acidic and basic reagents. The result of a typical experiment is shown in Fig. 9.2. Upon excitation at 262 nm the unmodified embedded film exhibited the fluorescence band of 8-MQ ($\lambda_{max} = 383$ nm, Fig. 9.2a). After treatment of the film with a solution of CF$_3$COOH in MeCN followed by drying, the fluorescence band of 8-MQ was replaced by a band with $\lambda_{max} = 465$ nm (Fig. 9.2b), showing that the quantitative conversion of 8-MQ into 8-MQ-H$^+$ within the plastic material had occurred. Successive exposure of the film to a MeCN solution of nBu$_3$N regenerated the fluorescence spectrum of 8-MQ (Fig. 9.2c), and a second switching cycle could be started (Fig. 9.2d). Therefore, our observations indicate that the protonation–deprotonation reactions of 8-MQ/8-MQ-H$^+$ also take place reversibly when such molecules are embedded in a polystyrene matrix (Fig. 9.3).

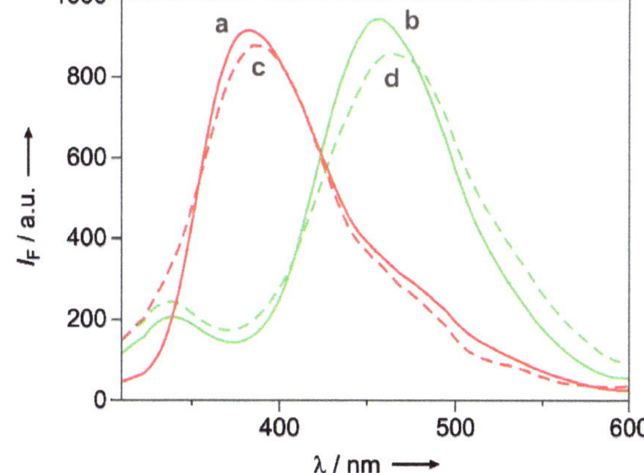

Fig. 9.2 Fluorescence spectra ($\lambda_{ex} = 262$ nm, RT) of a 0.7 μm-thick polystyrene film containing 8-MQ as prepared (*a*) and subjected to the following sequence of treatments: 10^{-2} mol L^{-1} solution of CF$_3$COOH in MeCN followed by drying (*b*), 10^{-2} mol L^{-1} solution of nBu$_3$N in MeCN followed by drying (*c*), 10^{-2} mol L^{-1} solution of CF$_3$COOH in MeCN followed by drying (*d*)

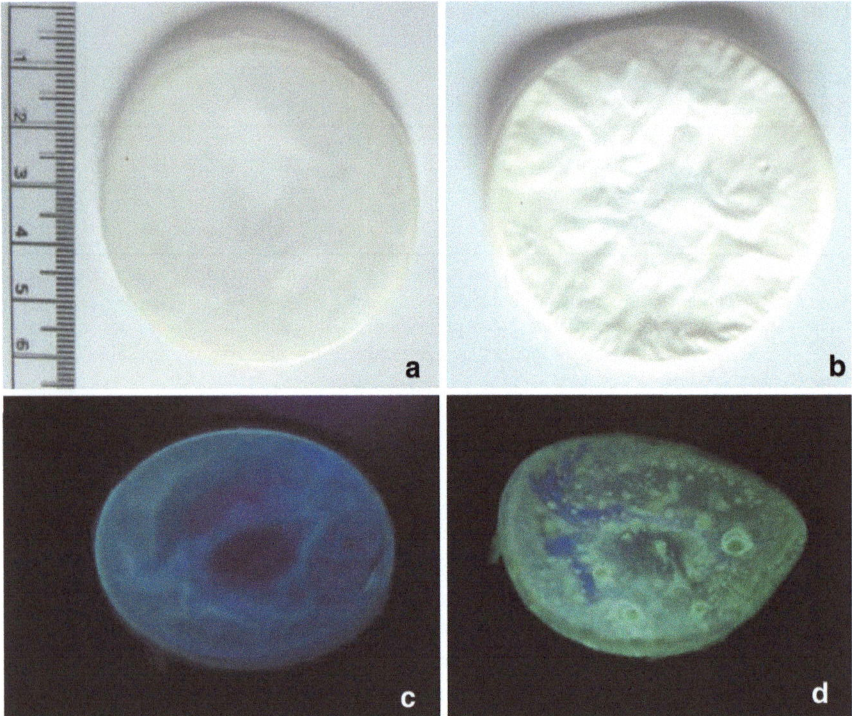

Fig. 9.3 Photographs of a polystyrene solid film containing 8-MQ under ambient light (*top*) and under irradiation with a near-UV lamp (*bottom*; λ_{irr} = 365 nm; a cutoff filter at 395 nm was placed in front of the camera to reduce the effect of scattered excitation light). Conditions: **a** and **c**, as-prepared film; **b** and **d**, after treatment with a MeCN solution of CF_3COOH on both sides followed by drying in air

9.3 Conclusion

In summary, we have shown that the proton-driven reversible modulation of two complementary absorption and fluorescence signals allows the implementation of both 2:1 multiplexer and 1:2 demultiplexer digital functions with an unsophisticated fluorophore. This work demonstrates that functions achieved by combinational circuits whose logic design requires the interconnection of several basic elements can be implemented with simple molecules, taking advantage of logic reconfiguration. Our next goal is to develop a system mimicking a real communication network, in which the output of a multiplexer is sent as the input to a separate demultiplexer. Observation of switching at the single-molecule level (for recent examples of single-molecule fluorescence switching, see: [42–46]) and control of the address input with light by intermolecular proton transfer from a photoacid [47, 48] in order to obtain complete optical operation [13] are also under investigation.

References

1. Goodsell DS (2004) Bionanotechnology—lessons from nature. Wiley, Hoboken
2. Balzani V, Credi A, Venturi M (2008) Chem Eur J 14:26
3. Lehn J-M (2002) Proc Natl Acad Sci USA 99:4763
4. Prasanna de Silva A, James DY, McKinney BOF, Pears DA, Weir SM (2006) Nat Mater 5:787
5. Rinaudo K, Bleris L, Maddamsetti R, Subramanian S, Weiss R, Benenson Y (2007) Nat Biotechnol 25:795
6. Feringa BL (ed) (2001) Molecular switches. Wiley-VCH, Weinheim
7. Gregg JR (1998) Ones and zeros: understanding boolean algebra digital circuits and the logic of sets. Wiley, New York
8. de Silva AP, Gunaratne HQN, McCoy CP (1993) Nature 364:42
9. Balzani V, Credi A, Venturi M (2008) Molecular devices and machines-concepts and perspectives for the nanoworld, 2nd edn. Wiley-VCH, Weinheim Chap. 9
10. Balzani V, Credi A, Venturi M (2003) ChemPhysChem 4:49
11. de Silva AP, Uchiyama S (2007) Nat Nanotechnol 2:399
12. Gust D, Moore TA, Moore AL (2006) Chem Commun 1169
13. Raymo FM, Tomasulo M (2006) Chem Eur J 12:3186
14. Pischel U (2007) Angew Chem 119:4100
15. Pischel U (2007) Angew Chem Int Ed 46:4026
16. Credi A (2007) Angew Chem 119:5568
17. Credi A (2007) Angew Chem Int Ed 46:5472
18. Langford SJ, Yann T (2003) J Am Chem Soc 125:11198
19. Qu DH, Wang QC, Tian H (2005) Angew Chem 117:5430
20. Qu DH, Wang QC, Tian H (2005) Angew Chem Int Ed 44:5296
21. Szacilowski K, Macyk W, Stochel G (2006) J Am Chem Soc 128:4550
22. Remacle F, Weinkauf R, Levine RD (2006) J Phys Chem A 110:177
23. Margulies D, Melman G, Shanzer A (2006) J Am Chem Soc 128:4865
24. Pischel U, Heller B (2008) New J Chem 32:395
25. Niazov T, Baron R, Katz E, Lioubashevski O, Willner I (2006) Proc Natl Acad Sci USA 103:17160
26. Lederman H, Macdonald J, Stefanovic D, Stojanovic MN (2006) Biochemistry 45:1194
27. Seelig G, Soloveichnik D, Zhang DY, Winfree E (2006) Science 314:1585
28. Frezza BM, Cockroft SL, Ghadiri MR (2007) J Am Chem Soc 129:14875
29. Andréasson J, Straight SD, Bandyopadhyay S, Mitchell RH, Moore TA, Moore AL, Gust D (2007) Angew Chem 119:976
30. Andréasson J, Straight SD, Bandyopadhyay S, Mitchell RH, Moore TA, Moore AL, Gust D (2007) Angew Chem Int Ed 46:958
31. Andréasson J, Straight SD, Bandyopadhyay S, Mitchell RH, Moore TA, Moore AL, Gust D (2007) J Phys Chem C 111:14274
32. Perez-Inestrosa E, Montenegro J-M, Collado D, Suau R (2008) Chem Commun 1085
33. Drezet A, Koller D, Hohenau A, Leitner A, Aussenegg FR, Krenn JR (2007) Nano Lett 7:1697
34. Macyk W, Stochel G, Szaciłowski K (2007) Chem Eur J 13:5676
35. Bronson RT, Montalti M, Prodi L, Zaccheroni N, Lamb RD, Dalley NK, Izatt RM, Bradshaw JS, Savage PB (2004) Tetrahedron 60:11139
36. Singh P, Kumar S (2006) Tetrahedron Lett 47:109
37. Zhang H, Wang Q-L, Jiang Y-B (2007) Tetrahedron Lett 48:3959
38. Goldman M, Wehry EL (1970) Anal Chem 42:1178
39. Schulman SG (1971) Anal Chem 43:285
40. Schulman SG, Capomacchia AC (1973) J Am Chem Soc 95:2763
41. Bardez E, Devol I, Larrey B, Valeur B (1997) J Phys Chem B 101:7786

42. Habuchi S, Ando R, Dedecker P, Verheijen W, Mizuno H, Miyawaki A, Hofkens J (2005) Proc Natl Acad Sci USA 102:9511
43. White SS, Li HT, Marsh RJ, Piper JD, Leonczek ND, Nicolaou N, Bain AJ, Ying LM, Klenerman D (2006) J Am Chem Soc 128:11423
44. Al-Kaysi RO, Bourdelande JL, Gallardo I, Guirado G, Hernando J (2007) Chem Eur J 13:7066
45. Kolaric B, Sliwa M, Brucale M, Vallée RAL, Zuccheri G, Samorì B, Hofkens J, De Schryver FC (2007) Photochem Photobiol Sci 6:614
46. Fukaminato T, Umemoto T, Iwata Y, Yokojima S, Yoneyama M, Nakamura S, Irie M (2007) J Am Chem Soc 129:5932
47. Raymo FM, Alvarado RJ, Giordani S, Cejas MA (2003) J Am Chem Soc 125:2361
48. Silvi S, Arduini A, Pochini A, Secchi A, Tomasulo M, Raymo FM, Baroncini M, Credi A (2007) J Am Chem Soc 129:13378